CAMBRIDGE LIBRARY COLLECTION

Books of enduring scholarly value

Technology

The focus of this series is engineering, broadly construed. It covers technological innovation from a range of periods and cultures, but centres on the technological achievements of the industrial era in the West, particularly in the nineteenth century, as understood by their contemporaries. Infrastructure is one major focus, covering the building of railways and canals, bridges and tunnels, land drainage, the laying of submarine cables, and the construction of docks and lighthouses. Other key topics include developments in industrial and manufacturing fields such as mining technology, the production of iron and steel, the use of steam power, and chemical processes such as photography and textile dyes.

The Creators of the Age of Steel

William Tulloch Jeans (1848–1907) was a parliamentary journalist with an interest in economics and technology. This book was first published in 1884, and comprises biographies of six men whom Jeans believed to have made significant contributions to the development of modern steel technology. The Bessemer process revolutionised steel-making, reducing the cost and allowing steel to replace the much more brittle iron in civil engineering projects such as bridges. Siemens' regenerative furnace allowed much more fuel-efficient manufacture of steel. Sir Joseph Whitworth developed a method of producing stronger steel by removing blowholes in the ingots. Sir John Brown's rolled steel was used in almost all the British navy's armour-plated ships. The work of Sidney Gilchrist Thomas and George Snelus on reducing phosphorus content in steel meant low-grade ores could be used. The combined researches of these men transformed modern industrial and engineering methods.

T0280213

Cambridge University Press has long been a pioneer in the reissuing of out-of-print titles from its own backlist, producing digital reprints of books that are still sought after by scholars and students but could not be reprinted economically using traditional technology. The Cambridge Library Collection extends this activity to a wider range of books which are still of importance to researchers and professionals, either for the source material they contain, or as landmarks in the history of their academic discipline.

Drawing from the world-renowned collections in the Cambridge University Library, and guided by the advice of experts in each subject area, Cambridge University Press is using state-of-the-art scanning machines in its own Printing House to capture the content of each book selected for inclusion. The files are processed to give a consistently clear, crisp image, and the books finished to the high quality standard for which the Press is recognised around the world. The latest print-on-demand technology ensures that the books will remain available indefinitely, and that orders for single or multiple copies can quickly be supplied.

The Cambridge Library Collection will bring back to life books of enduring scholarly value (including out-of-copyright works originally issued by other publishers) across a wide range of disciplines in the humanities and social sciences and in science and technology.

The Creators of
the Age of Steel

WILLIAM TULLOCH JEANS

CAMBRIDGE
UNIVERSITY PRESS

CAMBRIDGE UNIVERSITY PRESS

Cambridge, New York, Melbourne, Madrid, Cape Town, Singapore,
São Paolo, Delhi, Dubai, Tokyo, Mexico City

Published in the United States of America by Cambridge University Press, New York

www.cambridge.org
Information on this title: www.cambridge.org/9781108026925

© in this compilation Cambridge University Press 2011

This edition first published 1884
This digitally printed version 2011

ISBN 978-1-108-02692-5 Paperback

THE

CREATORS OF THE AGE OF STEEL.

THE CREATORS

OF

THE AGE OF STEEL.

BY

W. T. JEANS.

" Our wealth, commerce, and manufactures grow out of the skilled labour of men working in metals."—COBDEN.

LONDON: CHAPMAN AND HALL,
LIMITED.
1884.

CONTENTS.

CONTENTS.

CHAPTER IV.

CHAPTER V.

SIR WILLIAM SIEMENS.

CHAPTER VI.

CHAPTER VII.

SIR JOSEPH WHITWORTH.

CHAPTER VIII.

THE

CREATORS OF THE AGE OF STEEL.

INTRODUCTION.

So many characteristics are attributed to the present age that the title given to it in this work may appear to require some vindication. If this title is not " as true as steel," it has not been chosen without regard to precedent. When Professor J. D. Forbes wrote his *Dissertation on the Progress of Mathematical and Physical Science*, he began by stating that the year 1850 completed the third century of modern scientific progress, or the fourth if we include its dawn ; and that to each of these ages of discovery might be assigned a peculiarity in the character of its improvements, and even in the methods which conduced to that improvement. Taking the century from 1450 to 1550 as the dawn of physical science, the next century (1550—1650) he described as the first of true scientific activity, whose characteristic feature was the vindication of observation and experiment as the prime essentials to the increase of natural knowledge. Galileo was the glory of that age. The next hundred years (1650—1750) saw the triumphant application of mathematics to mechanics and physics. The transcendent genius of Newton illumined that age.

B

The distinguishing characteristic of the latter century (1750—1850) was the greatly extended use of experiment as a guide to truth, and the application of the knowledge thus acquired with more freedom and boldness for the benefit of mankind. It witnessed the development of the sciences of heat, electricity, and optics; and gave birth to the steam-engine of Watt, to the gigantic telescopes of Herschel and Lord Rosse, to the electric telegraph, and the tubular bridge. Watt and Faraday were its greatest adornments. The period which has elapsed since 1850 is that which we have designated "the age of steel"; for whatever other characteristics it may have, it certainly has been the birthtime of great inventions in metallurgy, and has witnessed unparalleled progress in those industries of which iron and steel form the essential mechanism. From a metallurgical as well as a chronological point of view the title is appropriate. Dr. Evans has justly remarked that such terms as iron age, bronze age, or stone age, mean only certain stages of civilisation, and not only chronological periods applicable to the whole of the world. Accordingly, the history of the manufacture of iron has hitherto been divided into two separate periods: the first extending from the earliest historical notice of it extant down to the time of the introduction of coal as the fuel used in smelting and refining iron; and the second extending from this date down to the introduction of the Bessemer process for converting iron into steel. When the latter was first announced to the world in 1856, it was described by its inventor as a mode of manufacturing malleable iron and steel *without fuel.* This was the dawn of the age of steel.

It is an age that has witnessed a great, though almost silent revolution in the manufacture of the most useful of all the metals. In 1851 Dr. Whewell remarked that "gold and iron at the present day, as in ancient times, are the rulers of the world; and the great events in the world of mineral art are not the discovery of new substances, but of new and rich localities of old ones—the opening of the treasures of the earth in Mexico and Peru in the sixteenth century, in California and Australia in our own day." Eleven years afterwards Mr. Cobden stated in the House of Commons that " our wealth, commerce, and manufactures grow out of the skilled labour of men working in metals." In the interval that elapsed between the making of these two statements, a single metallurgical invention was conceived, perfected, and put in operation, which, according to M. Michael Chevalier has been of more value than all the gold of California.[1] That invention has had a numerous offspring.

While science and mechanical skill have combined to produce wondrous results, said Sir Wm. Siemens, in 1869, " the germs of further and still greater achievements are matured in our mechanical workshops, in our forges, and in our metallurgical smelting works; it is there that the materials of construction are prepared, refined, and put into such forms as to render greater and still greater ends attainable. Here a great revolution of our constructive art has been prepared by the production in large quantities and at a moderate cost, of a material of more than twice the strength of iron, which, instead of being fibrous, has its full strength in every direction, and which can be modulated to every degree of ductility, approaching the hardness of the diamond on the one hand, and the proverbial toughness of leather on the other." This is the metal universally known as steel.

---

[1] This was no mere figure of speech. At the end of 1882 it was estimated that the probable yield of gold in the State of California, which is regarded as the best gold field yet discovered in America, was £230,000,000. The Bessemer process has saved even more than that vast sum.

This country has not only been the birthplace of all the great inventions which have increased or cheapened the production of iron and steel ; but of all the great industries carried on in this country the manufacture of iron and steel has made the greatest progress. Between 1851 and 1881 our cotton trade increased twofold, our coal trade threefold, our crude iron trade fourfold, and our steel trade thirtyfold. The production of steel in 1882 was as great as the production of crude iron in 1850.

Other industries have reflected this progress not only in their increasing magnitude, but in the greater use of mechanical appliances instead of manual labour. Just look at the cotton trade. England has long been the world's greatest cotton factory ; and although it is dependent on other countries for its raw material, this is the industry in which it has on the whole made the most uniform, though not always the most rapid progress. Thus, in 1847 we exported in round numbers one thousand million yards of cotton goods ; in 1857 two thousand million yards ; in 1867 three thousand million ; and in 1877 four thousand million. In other words, our exports of cotton goods have increased at the rate of one thousand million yards every ten years ; and to such an extent has machinery superseded manual labour in this industry, that on the average only one person is now employed to do the work that required two in 1850 ; while at the same time the hours of labour have been reduced, wages have been doubled, and the cost of manufacturing cotton goods has become cheaper. Similar results have been effected in other textile industries.

In like manner in our shipping trade, where iron or steel has almost entirely superseded wood, in 1881 only one man was employed where three were needed in 1850 ; while the number of hands only increased 36 per cent. the tonnage carried increased 360 per cent.

If this examination were extended to other leading industries in which machinery is largely used, it would be found that if this country were as much dependent on manual labour now as it was in 1850, our artizan population would have to be increased by two millions. In other words, if the additional work now done by machinery were done in the same way as in 1850, the additional number of artizans that would be required would have equalled in 1881 the total population of, Glasgow, Liverpool, Manchester, and Birmingham.

About one half of the whole of the world's coal trade, iron trade, cotton trade, and shipping trade is carried on in this country.

These few facts are calculated to show in a general way the part which iron and steel play in our industrial life. It is surely not too much to say that the iron and steel trade is the back-bone of our mechanical industries. Yet the nature of this "leading industry" is still little known to the masses of people who have never been in an iron or steel works. Although simple and inexpensive accounts of the processes employed in these works have been published of late years, they appear to have no fasci-nation for the popular mind. Be they ever so lucid, being colour-less and uneventful, they are not popular reading. It has therefore appeared to us that the most striking facts in connection with this industry could be presented in their most attractive light in short biographies of the creators of the age of steel. The career of men who with no birthright but their talents, and no other secret of success than "the magic of patience," have attained positions of world-wide renown, should always have an intrinsic interest; and biographies of such men, while affording scope for a narrative sufficiently varied by incident to sustain attention, may become the means of disseminating some knowledge of the great industry they have created.

In support of this view the high authority of Lord Bacon can be quoted. In order that " the industry of others may be quickened, and their courage aroused and inflamed," he remarked that " the introduction of great inventions appears to hold by far the first place among human actions, and it was considered so in former ages ; for to the authors of inventions they awarded divine honours, but only heroic honours to those who did good service in the state (such as the founders of cities and empires, legislators, deliverers of their country from long endured misfortunes, quellers of tyrannies, and the like). And certainly, if any one rightly compare the two, he will find that this judgment of antiquity was just ; for the benefits of inventions may extend to the whole race of man, but civil benefits only to particular places ; the latter, moreover, last not beyond a few ages, the former for ever. The reformation of the state in civil matters is seldom brought about without violence and confusion, while inventions carry blessings with them, and confer benefits without causing harm or sorrow to any."

Subservient to this, there are other purposes which such a narrative is calculated to promote. The industrial revolution described in the following pages has been brought about by one of those constellations of genius which the world has only seen at distant intervals, and which make their generation an epoch in the annals of science. But history teaches us that this sort of radiance has been ofttimes succeeded by an almost moonless night. It has been well observed, for instance, that the publication of the *Principia*, the crowning work of the creator of philosophical science, seemed to check the tide of invention in England, and that perhaps the splendour of Newton's fame had some influence for a season in over-dazzling the lustre of native talent. As a parallel instance, it has been remarked that nothing pre-eminent in science was produced by the French for half a century after Descartes. But this over-dazzling splendour has too often in the past had to pierce through the deepest shades of penury and neglect, which gave little encouragement to others to follow in the same path. Might not the next generation be spared from a similar eclipse by an account of the unprecedented success of the creators of the age of steel?

The rewards and honours of political life lead many men to seek after them; but there is a time-honoured tradition that great inventors and scientists, like poets and philosophers of old, are allowed to go unrewarded. The records of the past tell us, with a truthfulness that appears stranger than fiction, that many of the men who did most by their inventive faculties and life-long labours to ameliorate the condition of their fellow-men or to promote the industrial progress of their country, have ended their days in obscurity and poverty. The iron trade is not without its benefactors of this description. The two first and greatest inventors in the trade died in unenviable circumstances. Dud Dudley, who in 1618 discovered the way to smelt iron with coal, and thereby laid the foundation of many fortunes, concludes his record of his long and costly experiments by stating that "the author hath had no benefit thereby," and history tells us that he lived a life of hardship and died in obscurity. Henry Cort, who a hundred years ago invented the puddling process, which is still used for refining iron, was in his latter days a ruined man, and was only saved from absolute penury by a pension of £200 a year, granted by Mr. Pitt as a provision for himself, then aged fifty-four years, and his destitute family of twelve children. In a work written by the greatest authorities on the leading industries in 1851, we read: " The lot of the scientific man has hitherto been most frequently to expend years of study, experience, and research—his means, possibly his health; for what return? To find himself un-recognised, unheeded, and each year a poorer man than he was the year before." It is needless to multiply illustrations or quotations. Suffice it to say that the creators of the age of steel have been the pioneers of a happier era for the devotees of applied science; their success shows that the age which their labours inaugurated has been the dawn of more fortunate times for successful inventors; and the rewards and honours which they have received form not the least interesting portion of contemporary history.

During the last few years a good many inducements and facilities have been held out to young men with the view of attracting their attention to those departments of applied science or mechanical skill which have hitherto been less fashionable than professional life; and eminent authorities are daily pointing out the vast field there is for the study of science and its applications to industry. In this respect the iron and steel trade is no exception, notwithstanding the great achievements of the present generation. Sir Henry Bessemer stated at Sheffield in 1880 that to the young aspirant who would devote himself to the improvement of his art, it was above all things important that he should realise the fact that there lay before him a wide and open path for study and improvement, that its avenues were not in any way closed by what had already been done, and that in all things pertaining to the arts and manufactures we were as far removed from finality as the half-civilised Asiatic was still behind the European.

Next to the illimitable nature of the field there is for the exercise of genius, what is more likely to stimulate exertion in this direction than a knowledge of the rewards and honours gained by those who, although the foremost men of our day, still say that they have only gathered a few pebbles from the vast ocean that lay before them ? Admonitions to pursue science for the love of it have a celestial flavour about them ; but in matters terrestrial perhaps Diogenes showed as much knowledge of human nature as caustic wit, when in reply to the question " How it was that the philosophers followed the rich, instead of the rich following the philosophers," he answered— " Because the philosophers know what they want, and the rich do not."

The main object of biography, said Goethe, is to exhibit man in relation to the features of his time ; and this has been the guiding principle in the compilation of this volume. While it is not consistent with its purpose to give the full personal details that compose a *life*, it gives such facts and incidents as the public have a right to know, and such as the subjects themselves have thought proper to make public.

SIR HENRY BESSEMER.

CHAPTER I.

" The respect which in all ages and countries has been paid to inventors seems indeed to rest on something more profound than mere gratitude for the benefits which they have been the means of conferring on mankind, and to imply, if it does not express, a consciousness that by the grand, original conceptions of their minds they approach somewhat more nearly than their fellows to the qualities and pre-eminence of a higher order of being."—MUIRHEAD.

In the commercial history of the last hundred years there are three events that have had a revolutionary effect in accelerating our industrial development. The first was the construction of the steam-engine by James Watt, the second was the introduction of the penny post by Sir Rowland Hill, and the third was the invention of means of producing cheap steel by Sir Henry Bessemer and Sir Wm. Siemens. It is a remarkable feature of each of these great improvements that they came perfect from the hands of their authors. Of the steam-engine Sir Wm. Siemens has well remarked that if any proof were wanting of the great genius of Watt, it would be sufficient to observe that the steam-engine of the present day is in point of principle still the same as it left his hands three-quarters of a century ago, and that our age of material progress could only affect its form. Sir William Armstrong has likewise said that by a succession of brilliant inventions, comprising, amongst others, his parallel motion and his ball governor, Watt advanced to the final conception of the double-acting rotative engine, which became applicable to every purpose requiring motive power, and continues to this day, in nearly its original form, to be the chief moving agent employed by man. The two other inventions we have named might be described in similar terms. It is well known that Sir Rowland Hill not only conceived but perfected the organisation of the penny post; and the first announcement of the Bessemer converter, made in 1856, is still read with interest on account of its complete description of both the principle and details of that process. Each of these inventions had one other feature in common : they were brought into extensive use during the lifetime of their authors. In some respects, however, the Bessemer converter differed from the two other inventions. The improvement of the steam-engine was the only great invention of James Watt, and Sir Rowland Hill will be remembered for nothing but the penny post. Whether these inventions were accidental suggestions, happy thoughts, or flashes of genius, certain it is they were the only ideas of their authors that brought forth much fruit. Pope says :—

> " One science only will one genius fit,
> So vast is art, so narrow human wit."

But this cannot be said of our inventors in the steel trade. Their minds being of a more fertile stamp, have worked out many inventions, while at the same time they have realised the truth of Lord Bacon's axiom that knowledge to be profitable to its cultivators must also be fruitful to mankind. More especially has this been the case with the Bessemer converter, which has had the twofold effect of rewarding him who gave it and those to whom it was given. Sir Henry Bessemer was not only the architect of his own fortune but the benefactor of his race. The steam-engine and the penny post have in many respects been greater blessings to mankind; yet special Acts of Parliament were required to reward their authors for their labours James Watt wrote to the partner whose assistance saved him from ruin that " of all things in life there is nothing more foolish (unprofitable) than inventing." Sir Rowland Hill's first substantial reward was a public subscription, for the penny post was in operation sixteen years before it paid its own expenses. Nor can their life's work be described as inventions in every sense of the word. James Watt has been correctly described by Lord Jeffrey, his panegyrist, as the great improver of the steam-engine; and Sir Rowland Hill's great work was rather one of organisation than invention. But no such limitations have to be applied to the work of Sir Henry Bessemer. Biographers of inventors have often had great difficulty in vindicating the originality of their subjects from the claims of others to a previous knowledge or discovery of the same thing. James Watt, for instance, spent his life in improving the steam-engine and conducting law-suits to protect his patent rights; but the invention of Sir Henry Bessemer has in this country, the home of metallurgical inventions, been allowed to go unchallenged. It is related in Grecian history that after the battle of Salamis the generals, while each claiming the first honour for his own generalship, unanimously admitted that Themistocles deserved the second; and the world ever since, says Adam Smith, has accepted this as a proof that Themistocles was, beyond all question, the first. But no one in the steel trade has ever assigned to Sir Henry Bessemer a second place. His priority is undisputed. His inventive faculties were of such a high order that he boldly entered upon untrodden paths, and imparted features of originality to most things that engaged his attention.

In the year 1813, M. de Sismondi published his greatest work, in which he said that the French certainly possess, above every other nation of modern times, an inventive spirit. Such a remark, published two years before the battle of Waterloo, Englishmen would probably laugh at nowadays as the offspring of national vanity. Yet in that same year one of the most inventive spirits of his age first saw the light in circumstances that stand associated with one of the most eventful epochs in French history. At the time of the great revolution of 1792 there was employed in the French Mint a man of great ingenuity, who had become a member of the French Academy of Sciences at the age of twenty-five. When Robespierre became Dictator of France this scientific academician was transferred from the Mint to the management of a public bakery, established for the purpose of supplying the populace of Paris with bread. In that position he soon became the object of revolutionary frenzy. One day a rumour was set afloat that the loaves supplied were light in weight; and spreading like wildfire it was made the occasion of a fearful tumult. The manager of the bakery was instantly seized and cast into prison. He succeeded in escaping, but it was at the risk of his life. Knowing the peril he was in, he lost no time in making his way to England, and he only succeeded in doing so by adroitly using some documents he possessed bearing the signature of the Dictator. Landing in England a ruined man, his talents soon proved a passport to success. He was appointed to a situation in the English Mint; and by the exercise of his ingenuity in other directions, he ere long acquired sufficient means to buy a small estate at Charlton in Hertfordshire. Such, in brief, were the circumstances that led to the settlement there of Anthony Bessemer, the father of Sir Henry Bessemer. The latter may be said to have been born an inventor. His father was an inventor before him. After settling in England, his inventive ingenuity was displayed in making improvements in microscopes and in typefounding, and in the discovery of what his son has happily described as the true alchemy. The latter discovery, which he made about the beginning of the present century, was a source of considerable profit to him. It is generally known that when gold articles are made by the jewellers, there are various discolorations left on their surface by the process of manufacture; and in order to clear their surface, they are put into a solution of alum, salt, and saltpetre, which dissolves a large quantity of the copper that is used as an alloy. Anthony Bessemer discovered that this powerful acid not only dissolved the copper, but also dissolved a quantity of gold. He accordingly began to buy up this liquor, and as he was the only one who knew that it contained gold in solution, he had no difficulty in arranging for the purchase of it from all the manufacturers in London. From that liquor he succeeded in extracting gold in considerable quantities for many years. By some means that he kept secret (and the secret died with him), he deposited the particles of gold on the shavings of another metal, which being afterwards melted, left the pure gold in small quantities. Thirty years afterward the Messrs. Elkington invented the electrotype process, which had the same effect. Anthony Bessemer was also eminently successful as a typefounder. When in France, before the revolution of 1792, he cut a great many founts of type for Messrs. Firmin Didot, the celebrated French typefounders; and after his return to England, he betook himself, as a diversion, to typecutting for Mr. Henry Caslon, the celebrated English typefounder. He engraved an entire series, from pica to diamond, a work which occupied several years. The success of these types led to the establishment of the firm of Bessemer and Catherwood as typefounders, carrying on business at Charlton. The great improvement which Anthony Bessemer introduced into the art of typemaking was not so much in the engraving as in the composition of the metal. He discovered that an alloy of copper, tin, and bismuth was the most durable metal for type: and the working of this discovery was very successful in his hands. The secret of his success, however, he kept unknown to the trade. He knew that if it were suspected that the superiority of his type consisted in the composition of the metal, analysis would reveal it, and others would then be able to compete with him. So to divert attention from the real cause, he pointed out to the trade that the shape of his type was different, as the angle at which all the lines were produced from the surface was more obtuse in his type than in those of other manufacturers, at the same time contending that his type would wear longer. Other manufacturers ridiculed this account of Bessemer's type, but experience showed that it lasted nearly twice as long as other type. The business flourished for a dozen years under his direction, and during that period the real cause of its success was kept a secret. The process has since been rediscovered and patented. Such were some of the inventive efforts of the father of one of the greatest inventors of the present age.

The youngest son of Anthony Bessemer, Henry, was born at Charlton, in Hertfordshire, in 1813. His boyhood was spent in his native village, and while receiving the rudiments of an ordinary education in the neighbouring town of Hitchin, the leisure and retirement of rural life afforded ample time, though perhaps little inducement, for the display of the natural bent of his mind. Notwithstanding his scanty and imperfect mechanical appliances, his early years were devoted to the cultivation of his inventive faculties. His parents encouraged him in his youthful efforts. It is a pious wish, says Goethe, of all fathers to see what they have themselves failed to attain realised in their sons ; as if in this way they could live their lives over again and at least make a proper use of their early experience. Be this as it may, Henry's father perceived the superior powers of his mind while he was yet a boy ; and on the occasion of one of his visits to London purchased for him "one of those beautiful little five-inch foot-lathes made by Holtzapffel," and with this instrument the youth began to study the art of turning. Aided by a book, published by the same firm, that showed by many excellent illustrations the process of eccentric turning, he devoted himself to the study of the art with all the enthusiasm of genius and youth. At that time Jacob Perkins caused some sensation by his achievements in eccentric turning applied to the engraving of bank-notes ; and Henry Bessemer so much admired this man's work that he took the only little bit of turning he had to do to Perkins in order that, notwithstanding its unusually high price, he might have the privilege, as he esteemed it, of examining workmanship which he considered so beautiful and difficult. " I examined," says Sir Henry, "a vast number of beautiful specimens, and could not conceive how by any combination of mechanical appliances they could be produced until I came across a pattern where there was a false division.   Some accident to the machine had produced the deflection of a single line among many thousands, and I was thus enabled to learn in a moment the whole secret.   In three months from that time I produced all those patterns by the simplest possible means."

At the age of eighteen he came to London, " knowing no one," he says, "and myself unknown,—a mere cipher in a vast sea of human enterprise."   Here he worked as a modeller and designer with encouraging success. He engraved a large number of elegant and original designs on steel with a diamond point for patent medicine labels.   He got plenty of this sort of work to do, and was well paid for it.   In his boyhood his favourite amusement was the modelling of objects in clay ; and even in this primitive school of genius he worked with so much success that at the age of nineteen he exhibited one of his beautiful models at the Royal Academy, then held at Somerset House. Thus he soon began to make his way in the Metropolis, and in the course of the following year he was maturing some plans in connection with the production of stamps which he sanguinely hoped would lead him on to fortune. At that time the old form of stamps were in use that had been employed since the days of Queen Anne; and as they were easily transferred from old deeds to new ones, the Government lost a large amount annually by this surreptitious use of old stamps instead of new ones. The ordinary impressed or embossed stamps such as are now employed on bills of exchange, or impressed directly on skins of parchment, were liable to be entirely obliterated if exposed for some months to a damp atmosphere. A deed so exposed would at last appear as if unstamped, and would therefore become invalid. Special precautions were therefore observed in order to prevent this occurrence. It was the practice to gum small pieces of blue paper on the parchment, and to render it still more secure a strip of metal foil was passed through it, and another small piece of paper with the printed initials of the sovereign was gummed over the loose end of the foil at the back. The stamp was then impressed on the blue paper, which, unlike parchment, is incapable of losing the impression by exposure to a damp atmosphere. Experience showed, however, that by placing a little piece of moistened blotting-paper for a few hours over the paper the gum became so softened that the two pieces of paper and the slip of foil could be easily removed from an old deed, and then used for a new one. In this way stamps could be used a second and third time; and by thus utilising the expensive stamps on old deeds of partnerships that were dissolved, or leases that had expired, the public revenue lost thousands of pounds every year. Sir Charles Persley, of the Stamp Office, told Sir Henry Bessemer that the Government were probably defrauded of £100,000 per annum in that way. The young inventor at once set to work for the express purpose of devising a stamp that could not be used twice. His first discovery was a mode by which he could have reproduced easily and cheaply thousands of

c

stamps of any pattern. " The facility," he says, "with which I could make a permanent die from a thin paper original, capable of producing a thousand copies, would have opened a wide door for successful frauds if my process had been known to un-scrupulous persons, for there is not a government stamp or a paper seal of a corporate body that every common office clerk could not forge in a few minutes at the office of his employer or at his own home. The production of such a die from a common paper stamp is a work of only ten minutes; the materials cost less than one penny; no sort of technical skill is necessary, and a common copying press or a letter stamp yields most successful copies." To this day a successful forger has to employ a skilful die-sinker to make a good imitation in steel of the document he wishes to forge; but if such a method as that discovered and described by Sir Henry Bessemer were known, what a prospect it would open up! Appalled at the effect which the communication of such a process would have had upon the business of the Stamp Office, he carefully kept the knowledge of it to himself, and to this day it remains a profound secret.

More than ever impressed with the necessity for an improved form of stamp, and conscious of his own capability to produce it, he laboured for some months to accomplish his object, feeling sure that, if successful, he would be amply rewarded by the Government. To ensure the secrecy of his experiments, he worked at them during the night after his ordinary business of the day was over. He succeeded at last in making a stamp which obviated the great objection to the then existing form, inasmuch as it would be impossible to transfer it from one deed to another, to obliterate it by moisture, or to take an impression from it capable of producing a duplicate. Flushed with success and confident of the reward of his labours, he waited upon Sir Charles Persley at Somerset House, and showed him by numerous proofs how easily all the then existing stamps could be forged, and his new invention to prevent forgery. Sir Charles, who was much astonished at the one invention and pleased with the other, asked Sir Henry to call again in a few days. At the second interview Sir Charles asked him to work out the principle of the new stamping invention more fully. Accordingly Sir Henry devoted five or six weeks' more labour in perfecting his stamp, with which the Stamp Office authorities were now well pleased. The design, as described by the inventor, was circular, about $2\frac{1}{2}$ in. in diameter, and consisted of a garter with a motto in capital letters, surmounted by a crown. Within the garter was a shield with the words "Five pounds." The space between the shield and the garter was filled with network in imitation of lace. The die was executed in steel, which pierced the parchment with more than 400 holes, and these holes formed the stamp. It is by a similar process that valentine makers have since then learned to make the perforated paper used in their trade. Such a stamp removed all the objections to the old one. So pleased was Sir Charles with it that he recommended it to Lord Althorp, and it was soon adopted by the Stamp Office. At the same time Sir Henry was asked whether he would be satisfied with the position of Superintendent of Stamps with £500 or £600 per annum as compensation for his invention instead of a sum of money from the Treasury. This appointment he gladly agreed to accept, for being engaged to be married at the time, he thought his future position in life was settled. Shortly afterwards he called on the young lady to whom he was engaged and communicated the glad tidings to her, at the same time showing her the design of his new stamp. On explaining to her that its chief virtue was that the new stamps thus produced could not, like the old ones, be fraudulently used twice or thrice, she instantly suggested that if all stamps had a date put upon them they could not be used at a future time without detection. This idea was new to him, and impressed

C 2

with its practical character he at once conceived a plan for the insertion of movable dates in the die of his stamp. The method by which this is now done is too well known to require description here; but in 1833 it was a new invention. Having worked out the details of a stamp with movable dates, he saw that it was more simple and more easily worked than his elaborate die for perforating stamps; but he also saw that if he disclosed his latest invention it might interfere with his settled prospects in connection with the carrying out of his first one. It was not without regret, too, that he saw the results of many months of toil and the experiments of many lonely nights at once super-seded; but his conviction of the superiority of his latest design was so strong, and his own sense of honour and his confidence in that of the Government so unsuspecting, that he boldly went and placed the whole matter before Sir Charles Persley. Of course the new design was preferred. Sir Charles truly observed that with this new plan all the old dies, old presses, and old workmen could be employed. Among the other advantages it presented to the Government, it did not fail to strike Sir Charles that no superintendent of stamps would now be necessary—a recommendation which the perforating die did not possess. The Stamp Office therefore abandoned the first invention of Sir Henry Bessemer in favour of his latest one, which is still in use. At the same time the Government abandoned the ingenuous and in-genious inventor. The old stamps were called in and the new ones issued in a few weeks; the revenue from stamps grew enormously, and forged or feloniously used stamps are now almost unheard of. The Stamp Office reaped a benefit which it is scarcely possible to estimate fully, while Sir Henry did not receive a farthing. Shortly after the new stamp was adopted by Act of Parliament Lord Althorp resigned, and his successors dis-claimed all liability. When the disappointed inventor pressed his claim, he was met by all sorts of half-promises and excuses,

which ended in nothing. The disappointment was all the more
galling, because if Sir Henry had stuck to his first adopted
plan his services would have been indispensable to its execution ;
and it was therefore through his putting a better and more easily
worked plan before them that his services were coolly ignored.
" I had no patent to fall back upon," he says, in describing the
incident afterward; " I could not go to law even if I wished to
do so, for I was reminded when pressing for mere money out of
pocket that I had done all the work voluntarily and of my own
accord. Wearied and disgusted, I at last ceased to waste time
in calling at the Stamp Office—for time was precious to me in
those days—and I felt that nothing but increased exertions could
make up for the loss of some nine months of toil and expenditure
Thus, sad and dispirited, and with a burning sense of injustice
overpowering all other feelings, I went my way from the Stamp
Office too proud to ask as a favour that which was indubitably
my right."

Experience, says Carlyle, is the best schoolmaster, only the
school fees are rather heavy. Sir Henry Bessemer had now learned
this lesson. Success or failure in first efforts often mould the
course of after life. From a pecuniary point of view his experience
so far only appeared to confirm the old tradition that inventors are
doomed to disappointment ; but as it is one of the tests of genius
that it looks upon difficulties as things to be overcome, it is
scarcely wonderful that disappointments do not daunt it. On the
contrary adversity often acts as a stimulant. Necessity is the mother
of invention. Fortunately this was the case with Sir Henry
Bessemer. While smarting from the injustice of the English
government, he was encouraged by the mechanical success of the
invention they had appropriated. He therefore continued to work
out new inventions, but took care in future to turn them to more
profitable account and to protect them from piracy.

About the time that he was busily but unprofitably engaged in frustrating the fraudulant use of government stamps, his attention was called to the difficulty of obtaining good patterns of figured Utrecht velvet; and he soon invented a machine that overcame this difficulty.   It was so successful in operation that some of the velvet it produced was used in furnishing certain state apartments in Windsor Castle.   Several of the designs in figured velvet still in use were produced by our young inventor.

The next matter that seriously engaged his attention was the process of typecasting, improvements in which formed the subject of his first patent.   In his youth he had made himself acquainted with the details of typefounding in his father's foundry at Charlton ; and he now designed new apparatus for casting type that contained many of the elements of the present typecasting machines.   His machine, which was patented on the 8th of March, 1838, produced the most accurate type ever cast up to that date, and by exhausting the air from the mould gave perfectly solid bodies ; but the little valve through which the metal was injected into the mould used to fail after casting some six or eight thousand types, and owing to this defect the machine was eventually abandoned.   Some years afterwards he constructed what was known as Young's composing machine, with which the *Family Herald* was " composed " for about two years by a young lady, who with it could set from six to seven thousand types per hour for ten consecutive hours ; but ultimately the great opposition of the compositors led to its abandonment.

Shortly after he had taken out his first patent for his improve-ment in typefounding his attention was accidentally turned to the manufacture of bronze powder which is used in gold work, japanning, gold printing, and similar operations.   While engaged in ornamenting a vignette in his sister's album, he had to purchase a small quantity of this bronze and was struck with the great difference between the price of the raw material and that of the manufactured article.   The latter sold for 112s. a pound, while the raw material only cost 11*d.* a pound. He concluded that the difference was caused by the process of manufacture, and made inquiries with the view of learning the nature of the process. He found, however, that this manufacture was hardly known in England. The article was supplied to English dealers from Nuremburg and other towns in Germany. He did not succeed, therefore, in finding any one who could tell him how it was produced. In these circumstances he determined to try to make it himself, and worked for a year and a half at the solution of this task. Other men had tried it and failed, and he was on the point of failing too. After eighteen months of fruitless labour he came to the conclusion that he could not make it, and gave it up. But it is the highest attribute of genius to succeed where others fail, and impelled by this instinct he resumed his investigations after six months' repose. At last success crowned his efforts. The profits of his previous inventions now supplied him with funds sufficient to provide the mechanical appliances he had designed.

Knowing very little of the patent law and considering it so insecure that the safest way to reap the full benefit of his new invention was to keep it to himself, he determined to work his process of bronze making in strict secrecy; and every precaution was therefore adopted for this purpose. He first put up a small apparatus with his own hands, and worked it entirely himself. By this means he produced the required article at 4*s.* a pound. He then sent out a traveller with samples of it, and the first order he got was at 80*s.* a pound. Being thus fully assured of success, he communicated his plans to a friend who agreed to put 10,000*l.* into the business, as a sleeping partner, in order to work the new manufacture on a larger scale. The entire working of the concern was left in the hands of Sir Henry, who accordingly proceeded to enlarge his means of production. To ensure secrecy he made plans of all the machinery required, and then divided them into sections. He next sent these sectional drawings to different engineering works in order to get his machinery made piecemeal in different parts of England. This done, he collected the various pieces, and fitted them up himself a work that occupied him nine months. Finding everything at last in perfect working order, he engaged four or five assistants in whom he had confidence, and paid them very high wages on condition that they kept everything in the strictest secrecy. Bronze powder was now produced in large quantities by means of five different self-acting machines, which not only superseded hand labour almost entirely, but were capable of producing as much daily as sixty skilled operatives could do by the old hand system.

To this day the mechanical means by which his famous gold paint is produced remains a secret. The machinery is driven by a steam-engine in an adjoining room; and into the room where the automatic manufactory is at work none but the inventor and his assistants have ever entered. When a sufficient quantity of work is done a bell is rung to give notice to the engineman to stop the engine, and in this way the machinery has been in constant use for over forty years without having been either patented or pirated. Its profit was as great as its success. At first he made 1,000 per cent. profit; and though there are other products that now compete with this bronze, it still yields 300 per cent. profit. "All this time," says the successful inventor thirty years afterward, "I have been afraid to improve the machinery, or to introduce other engineers into the works to improve them. Strange to say we have thus among us a manufacture wholly unimproved for thirty years. I do not believe there is another instance of such a thing in the kingdom. I believe that if I had patented it the fourteen years would not have run out without other people making improvements in the manufacture. Of the five machines I use, three are applicable to other processes, one to colour making especially; so much so that notwithstanding the very excellent income which I derive from the manufacture, I had once nearly made up my mind to throw it open, and make it public for the purpose of using part of my invention for the manufacture of colours. Three out of my five assistants have died, and if the other two were to die and myself too, no one would know what the invention is." Since this was said, in 1871, Sir Henry has rewarded the faithfulness of his two surviving assistants by handing over to them the business and factory.

Among the other matters that engaged his attention between 1844 and 1854 were the manufacture of paints, oils, and varnishes; the manufacture of sugar, which formed the subject of several patents; the construction of railway carriages, centrifugal pumps, projectiles, and ordnance. Most of these inventions were so purely mechanical as scarcely to admit of description in a way that would be of popular interest now; and on the other hand some which were then recognised as valuable improvements have since been improved out of existence. In reading accounts of their introduction and use in contemporary publications, it is worthy of remark that their author is generally designated as " the ingenious Mr. Bessemer"—a tribute which his inventions had already earned for him in popular estimation. Some incidents connected with these inventions ought to be recorded.

In 1847 when he patented some improvements in railway carriages he wrote a pamphlet on the resistance of the atmosphere to railway trains. At that time it was the practice of railway companies to heap up luggage on the roofs of their carriages; and Sir Henry's pamphlet demonstrated the folly of this arrangement. He showed that a portmanteau measuring three feet by one on the top of a railway carriage going at the rate of thirty miles an hour presented a resistance of 13·5 lbs.; and assuming that 10 lbs. would draw a ton there was a resistance caused by the portmanteau equal to a weight of 2,880 lbs. inside the carriage: while if this rule were applied to express trains moving at sixty miles per hour, the resistance increasing as the square of the velocity, it would be equal to a load of five tons inside the carriage.   At that time not only was luggage piled up on the top of carriages, but the carriages themselves were placed wider apart from each other—two practices which were subsequently abandoned.

In 1849 a committee of the House of Commons was appointed to inquire into the causes of explosions in coal mines.   Sir. Henry thought this a favourable opportunity for bringing forward a plan for ventilating mines which he had patented five years previously. Accordingly he instantly had a working model of it constructed on his premises.   Two shafts were made and connected with each other by an underground tunnel, each being $3\frac{1}{2}$ feet in diameter and lined with brick and cement.   One shaft was left open and was intended to represent the down shaft of a mine, while to the other, or upcast shaft, was affixed the new ventilating apparatus which was capable of drawing out 15,000 cubic feet of air per minute, and was driven by an engine of about four-horse power.   This subterranean structure and its ventilator, which were completed in a fortnight, answered its purpose, and being thus ready to demonstrate its success, Sir Henry offered to Lord Wharncliffe, the chairman of the select committee, to give evidence on the subject ; but his offer was refused on the pretext that his invention had not been actually applied in any coal mine.

At the Great Exhibition of 1851 he exhibited four different kinds of machines which were considered ingenious combinations of simplicity and power.   One was a pump for land and sewer drainage, described as capable of discharging twenty tons of water per minute, and of draining in an hour one acre of land one foot deep in water.   Another machine, for separating molasses from crystal sugar, was represented as capable of doing as much work with one man as could be done by two men and two of the machines previously in use.   It was the only machine of its kind that received a prize medal.   A novel machine for grinding and polishing plate-glass was much admired. In it a slate table, on which the plate-glass was laid, had a series of grooves, and by extracting the air from these grooves by means of an air-pump a vacuum was formed, so that the pressure of the atmosphere on the upper side of the glass held it firmly on the table while it was being ground and polished. By turning a cock which admitted the air again, the plate of glass could be instantly removed. The plan in general use then for holding down sheets of glass was to imbed them in plaster-of-Paris—an operation which had to be performed four times for each plate, and in which no less than forty tons of plaster per week were consumed in some establishments. In the "Lectures on the Results of the Great Exhibition," delivered before the Society of Arts, Mr. Henry Hensman, in dealing with the department of civil engineering and machinery, makes prominent mention of this "valuable apparatus" as one of the most conspicuous improvements among the machines for working in mineral substances. That lecture was delivered in February, 1852. Sir Henry Bessemer was then a stranger to mineralogy; he had not yet entered upon the field of discovery in which he was destined to attain a colossal fortune and universal fame. Towards this consummation the events of the next three years unconsciously led him on.

In 1853, when the public mind was much exercised by the prospects of the impending war with Russia, Sir Henry Bessemer enthusiastically devoted his attention to the improvement of the projectiles and ordnance then in use. He soon constructed elongated projectiles to which a rotatory motion was imparted during their passage through the air without the aid of the rifled grooves which still continue to be made in our ordnance, and without any deviation from the true cylindrical bore of the gun. To effect this he made small passages lengthwise through the projectile, and open at the end nearest the breech of the gun. Through these passages a part of the exploded powder found its way, and being emitted from the opposite sides of the projectile, the reactive force of the exploded gunpowder produced the rotatory motion required. Among other peculiarities of this gun was an enlarged powder chamber—an improvement that was made the subject of experiment by other inventors a quarter of a century afterwards—and he consequently insisted on increasing the strength of his gun and the metal near the breech. To prevent inconvenience from this increased weight he constructed his guns in parts, which were bolted together. "By this mode of forming guns in various parts I am enabled," he said, "to use iron and steel in some cases, and thus form a gun of great strength, the parts of which are of comparatively little weight, while it also admits of the various parts being made of the metals most suitable to resist the peculiar strain and wear to which they are severely subjected when in use."

Seeing that the English government had no good artillery suitable for firing elongated projectiles, and considering the system of rifling grooves as dangerous, he made a series of experiments in his own grounds with six pounder shots, with which he got what he considered more than ample rotation in a smooth-bore gun. He then submitted his plan to the government authorities at Woolwich Arsenal, but it was simply pooh-poohed. They never tried it.

Shortly after this, while Sir Henry was on a visit to Paris with Lord John Hay, he attended a dinner given to General Hamlin and other distinguished French officers before their departure for the Crimea. At that dinner Sir Henry met Prince Napoleon, to whom he took occasion to mention his plan of firing elongated projectiles. So favourably was the prince impressed with the invention that he asked Sir Henry to explain it to the Emperor, and with this view arranged an interview. The Emperor was also pleased with the account Sir Henry was able to give of his success, and invited him to continue his experiments at Vincennes. But other business soon recalling Sir Henry to London, he went and asked the Emperor's permission to make his experiments in London and to bring the projectiles to Paris for trial. His Majesty not only consented, but at the close of the audience said : " In this case you will be put to some expense, but I will have that seen to." Sir Henry returned to London, and a few days afterwards received a letter from the Duc de Bassano, together with an autograph note from the Emperor, authorising him to draw on Baring Brothers, of London, for the cost of manufacturing projectiles, but leaving him to fill in any amount.

The experiments were accordingly continued in London. A good many projectiles were made and sent to the Polygon at Vincennes for trial. Two days before Christmas, when the ground was covered with six inches of snow, several thirty-pounder projectiles were fired through ten boarded targets standing in a straight line, each target being about 100 yards distant from the other. In this way it was shown by the circular holes made in these targets that the plan of the inventor imparted sufficient rotation to his elongated projectiles, which generally passed through seven of the targets. A mechanical device was also affixed to the mouth of the gun to show the precise amount of rotation by marking the projectiles ; and several shots recovered from the snow indicated from one and a half to two and a quarter rotations in passing through the length of the gun, being a greater twist than that produced by the ordinary system of rifling. These promising results were considered very satisfactory by the French authorities, and they fully justified the anticipations of their designer ; but just at the moment when success appeared to be on the point of crowning his labours an incident occurred that changed the whole course of his future life, that materially affected the industrial progress of the world, and afforded another illustration of the saying—What great events from little causes spring !

CHAPTER II.

"It was a just answer of Solon to Crœsus, who showed him all his treasures
—'Yes, sir, but if another should come with better iron than you, he would
be master of all this gold.' "—BACON.

It is but rarely, says Professor John Playfair, that we can lay
hold with certainty of the thread by which genius has been guided
in its first discoveries. This desideratum, however, is not wanting
in the case of the great invention that revolutionised the steel
trade. When Sir Henry Bessemer had shown to the French
military authorities at Vincennes the results of his system of
firing elongated projectiles from a light cast-iron smooth-bore gun,
Commander Minie, who superintended the trials, remarked to
him : " The shots rotate properly, but if you cannot get stronger
metal for your guns, such heavy projectiles will be of little use."
It was this observation that first led Sir Henry to think of the
possibility of improving the manufacture of iron. It suggested
to him a new field of invention, and he instantly determined

> To brave the perils that environ
> The man who dabbles in cast iron.

In reporting the results of his artillery experiments to the Emperor
Napoleon he intimated his intention of extending his researches
to the kinds of metal most suitable for artillery purposes. His
Majesty gave every encouragement to this new project, and

requested that the results might be communicated to him. With this intention Sir Henry left Paris for London.

"What I admire in Christopher Columbus," said Turgot, "is not that he discovered the new world, but that he went to look for it on the faith of an idea." But when Sir Henry Bessemer determined to make improvements in the manufacture of iron and steel he had not the least idea of how he was going to do it. Both the rudiments and the history of metallurgy were unknown to him, and at first sight no subject could appear less inviting. The process then in use for making steel had been practised for nearly a century without any improvement, and the history of its invention was by no means encouraging. An honest and skilful clockmaker named Huntsman, who lived at Doncaster in 1738, was so annoyed at the defective nature of the watch springs then used in his trade that he began at an early age to make experiments with the view of producing a better quality of steel. In 1740 he removed to the little village of Handsworth, a few miles from Sheffield, where he carried on his experiments night and day for some years. Little is known of the character of these experiments except that they were laborious, for he kept them strictly secret; but the fact that they were long continued before they were successful bespeaks many failures. Eventually, however, he did succeed in his aim, and the process which he then invented was the only one in use for the next hundred years. Till then the finest steel used in this country was made by the Hindoos, and the price of it, previous to Huntsman's invention, is said to have been about £10,000 a ton. The process of Huntsman produced equally good steel at prices ranging from £100 to £50 a ton.

Huntsman determined to keep his valuable invention a secret, and to work it exclusively for his own benefit. In 1770 he established a manufactory at Attercliffe, near Sheffield, and the men who worked in it were sworn to secrecy. At first his steel was defamed by the Sheffield manufacturers, but foreign consumers, who were less prejudiced, soon recognised its excellence. Ere long its fame became world wide. The Sheffield manufacturers then began to wonder how ever it was made ; and as the demand for it became great, they multiplied their efforts to fathom the secret which had been so well kept. At last it got out, and the stratagem by which it was obtained became anything but a secret. There are Sheffield manufacturers still living who have preserved the tradition of its disclosure. In the dismal darkness and bitter cold of a winter's night a pitiable-looking beggar knocked at the entrance to Huntsman's works, while snow was falling heavily and all outside was enveloped in repulsive gloom. The shivering beggar abjectly asked for warm shelter. The workmen instantly assented, and assigned him a corner of the building where he could rest and be warmed. He appeared drowsy with fatigue, and soon fell asleep. But it was a cat's sleep. While the unsuspecting workmen busily proceeded with their work, the sleeping beggar was " eying " them all the time ; and, as the process lasted several hours, he continued his feigned sleep for several hours too. It afterwards transpired that the begging impostor was an iron-founder who resided at Greenside, near Sheffield, and the success of his stratagem was soon attested by the erection in a few months of rival steel works similar to those of Huntsman.

It was this system of manufacture that was exclusively employed when Sir Henry's attention was directed to metallurgy. The iron then used for making steel was mostly imported from Sweden, Russia, and Madras, and small quantities of similar quality were manufactured near Ulverston, in North Lancashire. The bars of ordinary-looking iron were first subjected to a process of cementation in a furnace constructed for the purpose. In the centre of the furnace were placed two chests or troughs, which were heated all round by holes in the flues. What was technically called cement was simply the dust of charcoal made from hard woods.

Sometimes, however, soot was employed. The cement was sifted into the troughs to the depth of two inches; the bars were put in upon their narrow edges, with a space of about half an inch between each; powdered charcoal was next sifted in to the depth of an inch, and then a second set of bars were made to fit in between the first. In this way the chests or troughs, capable of holding from eight to twelve tons of bar iron, were filled to within six inches of the top. Above the iron was placed a covering of cement-powder and damp sand, in order thoroughly to exclude the air. The trough being then closed, the fire was applied for a few days until the iron had absorbed sufficient carbon to produce the kind of steel required. To produce certain qualities the bars were exposed to two or three successive cementations. Through a small hole made for the purpose a bar could at any time be drawn out to show the progress of the cementation, which was considered complete when the surface of the bars was covered with blisters. Hence this is known as *blistered steel*. Blisters, however, do not form its only new feature. The metal thus produced has different properties from those of the iron bar; its texture, instead of being fibrous, is granular or crystalline; the colour of the fracture is white, like frosted silver, and the crystals are large in proportion to the amount of carbon absorbed. In this condition it is generally unfit for forging. For this purpose it has to be condensed and made uniform by what is known as the process of shearing, so called from having been first used for the manufacture of shears for cutting the wool off sheep. In this process the bar of blistered steel was broken into lengths of about eighteen inches. Four or more of these lengths were bound into a bundle or faggot by means of a ring, to which a long handle was attached. This faggot was raised to a welding heat, and in that state subjected to he action of a large hammer, called a tilting hammer, which worked at the rate of from 150 to 360 strokes per minute. The different pieces then became united, and although the solid rod

D

thus formed soon ceased to be red-hot, the rapid blows of the hammer revived the redness, not the least wonderful feature of the process being to see the metal ignite under the action of the strokes. This treatment made the steel fit to be forged with the hammer into shears, edge tools, and cutting instruments. It is called *shear steel*.

This was the process by which the steel was manufactured that Huntsman succeeded in improving. His process produced cast steel, which is still described by Sheffield manufacturers as the finest of all makes. In this process the bars of blistered steel were broken into fragments, melted in crucibles, and cast into ingots. The crucibles were made of special kinds of clay, a small quantity of coke dust, and fragments of old pots. These materials being mixed with water and worked up, were then kneaded on the floor for five or six hours by the naked feet of men, who. kept incessantly trampling it, folding it over, and treading it again and again. Of this composition small crucibles or jars, two feet high, were made in a mould of cast iron, dried in a heated vault, and afterwards annealed at a red heat. They were lined with ganister, then considered the most refractory material that could be used for resisting intense heat. These crucibles were placed in a furnace, and then each one was filled with about thirty pounds of steel in fragments, together with a small quantity of powdered charcoal. Thus charged, they were well covered in, and the furnace was well supplied with fuel. Four hours of this intense heat generally made the steel sufficiently fluid for casting. Moulds being prepared for this purpose, each crucible was in due course picked out of the furnace with a pair of tongs. The cover was then removed, and the fluid metal poured or cast into the mould. It flowed like water, its colour being a white heat, tinged with blue ; and as the stream fell from the crucible to the mould the action of the air on minute portions caused them to burn with brilliant scintillations. This process of making steel was considered the most perfect during the period of nearly a hundred years that elapsed between the time that it was invented by Huntsman and the time it was studied by Sir Henry Bessemer.

At that time Sir Henry had no connection with the iron or steel trade, and knew little or nothing of metallurgy. But this fact he has always represented as being rather an advantage than a drawback. "I find," he says, "in my experience with regard to inventions, that the most intelligent manufacturers invent many small improvements in various departments of their manufactures, but generally speaking these are only small ameliorations based on the nature of the operation they are daily pursuing, while, on the contrary, persons wholly unconnected with particular businesses, are the men who make all the great inventions of the age. I find that persons wholly unconnected with any particular business have their minds so free and untrammelled to view things as they are, and as they would present themselves to an independent observer, that they are the men who eventually produce the greatest changes." It was in this spirit that he began his investigations in metallurgy. His first business was to make himself acquainted with the information contained in the best works then published on the subject. He also endeavoured to add some practical knowledge to what he had learned from books. With this view he visited the iron-making districts in the north, and there obtained an insight into the working merits and defects of the processes then in use. On his return to London he arranged for the use of an old factory in St. Pancras where he began his own series of experiments. He converted the factory into a small experimental ironworks, in which his first object was to improve the quality of iron. For this purpose he made many costly experiments without the desired measure of success, but not without making some progress in the right direction. After twelve months spent in these experiments he produced an improved quality of cast iron, which was almost as white as steel, and was both tougher and stronger than the best cast iron then used for ordnance. Of this metal he cast a small model gun, which was turned and bored. This gun he took to Paris and presented it personally to the Emperor as the result of his labours thus far. His Majesty encouraged him to continue his experiments, and desired to be further imformed of the results.

As Sir Henry continued his labours he extended their scope from the production of refined iron to that of steel; and in order to protect himself he took out a patent for each successive improvement. One idea after another was put to the test of experiment; one furnace after another was pulled down, and numerous mechanical appliances were designed and tried in practice. During these experiments he specified a multitude of improvements in the crucible process of making steel; but he still felt that much remained to be done. At the end of eighteen months, he says, "the idea struck me" of rendering cast iron malleable by the introduction of atmospheric air into the fluid metal. His first experiment to test this idea was made in a crucible in the laboratory. He there found that by blowing air into the molten metal in the crucible by means of a movable blow-pipe he could convert ten pounds or twelve pounds of crude iron into the softest malleable iron. The samples thus produced were so satisfactory in all their mechanical tests that he brought them under the notice of Colonel Eardley Wilmot, the then Super-intendent of the Royal Gun Factories, who expressed himself delighted and astonished at the result, and who offered him facilities for experimenting in Woolwich Arsenal. These facilities were extended to him in the laboratory by Professor Abel, who made numberless analyses of the material as he progressed with his experiments. The testing department was also put at his disposal for testing the tensile strength and elasticity of different

samples of soft malleable iron and steel. The first piece that
was rolled at Woolwich was preserved by Sir Henry as a memento.
It was a small bar of metal about a foot long and an inch wide ;
and was converted from a state of pig iron into malleable iron
in a crucible of only ten pounds. That small piece of bar after
being rolled was tried to see how far it was capable of welding ;
and he was surprised to see how easily it answered the severest
tests. After that he commenced experiments on a larger scale.
He had proved in the laboratory that the principle of purifying pig iron by atmospheric air was possible ; but he feared,
from what he knew of iron metallurgy, that as he approached
the condition of pure, soft malleable iron he must of necessity require a temperature that he could not hope to attain
under these conditions. In order to produce larger quantities
of metal in this way, one of his first ideas was to apply the air
to the molten iron in crucibles, and accordingly in October, 1855,
he took out a patent embodying this idea. He proposed to erect
a large circular furnace with openings for the reception of
melting pots containing fluid iron, and pipes were made to
conduct air into the centre of each pot and to force it among
the particles of metal. Having thus tested the purifying effect
of cold air introduced into the molten iron in pots, he laboured
for three months in trying to overcome the mechanical difficulties
experienced in this complicated arrangement. He wondered
whether it would not be possible to dispense with the pipes
and pots, and perform the whole operation in one large circular
or egg-shaped vessel. The chief difficulty in doing so was how
to force the air all through the mass of liquid metal. While this
difficulty was revolving in his mind, the labour and anxiety
entailed by previous experiments brought on a short but severe
illness ; and while he was lying in bed pondering for hours
upon the prospects of succeeding in another experiment with
the pipes and pots, it occurred to him that the difficulty might

be got over by introducing air into a large vessel from below into the molten mass within.

Though he entertained grave doubts as to the practicability of carrying out this idea, chiefly owing to the high temperature required to maintain the iron in a state of fluidity while the impurities were being burned out, he determined to put it to a working test, and on recovering health he immediately began to design apparatus for this purpose. He constructed a circular vessel measuring three feet in diameter and five feet in height, and capable of holding 7 cwt. of iron; and he next ordered a small powerful air-engine and a quantity of crude iron to be put down on the premises in St. Pancras that he had hired for carrying on his experiments. The name of these premises was Baxter House —formerly the residence of old Richard Baxter; and the simple experiment we are now going to describe has made that house for ever famous. The primitive apparatus being ready, the engine was made to force streams of air under high-pressure through the bottom of the vessel, which was lined with fire-clay, and the stoker was told to pour the metal when it was sufficiently melted in at the top of it. A cast iron plate—one of those lids which commonly cover the coal-holes in the pavement—was hung over the converter; and all being got ready, the stoker in some bewilderment poured in the metal. Instantly out came a volcanic eruption of such dazzling coruscations as had never been seen before. The dangling pot-lid dissolved in the gleaming volume of flame, and the chain by which it hung grew red and then white as the various stages of the process were unfolded to the gaze of the wondering spectators. The air-cock to regulate the blast was beside the converting vessel, and no one dared to go near it, much less to deliberately shut it. In this dilemma, however, they were soon relieved by finding that the process of decarburisation or combustion had expended all its fury; and, most wonderful of all, the result was steel! The new metal was tried. Its quality was good. The problem was solved. The new process appeared successful. The inventor was elated.

Astonished at his own success, he went to the Patent Office and examined all the patents to see whether anybody had done the same thing before. He could find no trace of such an operation, but observed that steam had been used in that way.[1] On seeing that, he also tried steam, but found that it did not answer like air. Nevertheless he specified both air and steam in his patent lest the patentees of the steam process might afterwards claim that their apparatus was sufficient for working the new process.

" The result of my first experiment," he says, " showed me that the highest temperature ever known in the arts could be produced by the simple introduction of atmospheric air into cast iron.

[1] From the autobiography of Mr. James Nasmyth, published in 1883, one might gather the impression that Mr. Nasmyth was sure to have discovered the Bessemer process, and that Sir Henry only anticipated him by a few days or weeks. But Mr. Nasmyth gave a very different account of the matter in 1871, when he said : " It is a very curious fact that Mr. Bessemer, on his first production of a specimen of the result of his process at Cheltenham, selected me out at the meeting of the British Association, in the reception room, where he said, ' Now, Nasmyth, you are the first man who should see the result of this, because I have founded it on an idea of your own ; your patent for steam puddling led me to this process.' Now my patent for steam puddling was a system of oxidising the carbon in cast iron by introducing steam *beneath the surface of the molten metal* so as to convert it into malleable iron." It was that system of introducing an oxidising agent beneath the surface of the molten metal that he said led him to think of this mode of carrying out his invention. The patent of Mr. Nasmyth for the use of steam applied from *below* was dated May 4, 1854 ; but previous patents for the use of steam from *above* were taken out by A. M. Perkins in 1843, and by R. Plant in 1849. All of them were unsuccessful. The first patent of Sir H. Bessemer in which air is mentioned as the oxidising agent is dated 17th October, 1855, and other three months were spent in experimenting before the idea of introducing the air from the bottom of a large converter struck him. The patent embodying the latter idea is dated February 11th, 1856.

That temperature was much more than sufficient to keep the malleable iron in a fluid state. After the experiments had been going on for six or seven months, and after having, in conjunction with my partner, Mr. Robert Longsden, spent £3,000 or £4,000 in experiments, and diverted my attention from business pursuits for about two years and a half, I was anxious to get some other opinion on the process, and I invited the late Mr. George Rennie to inspect it at my works."

On seeing the result of "a blow" in the converter, Mr. George Rennie said : "Whatever your difficulties are in working details, the moment a practical ironmaster sees this wonderful invention he will at once supply all those details, and the thing will be done. This must not be hid under a bushel. The British Association meets next week at Cheltenham ; if you have patented your invention draw up an account of it in a paper, and have it read in Section G." Acting on this suggestion, Sir Henry wrote out a description of his new invention, entitling it " The Manufacture of malleable Iron and Steel without Fuel." (About four tons of hard coke, equal to eight tons of coal, were then required to make a ton of cast steel.) He accordingly went down to Cheltenham, and was breakfasting next morning in the coffee-room of the Queen's Hotel with Mr. Clay, of the Mersey Steel Works, when a gentleman who did not know Sir Henry said : "Clay, I want you to go with me this morning.   There is a fellow who has come down from London to read a paper on making steel from cast-iron without fuel !   Ha ! ha ! ha !"   Mr. Clay consented to go, and in an hour or so the three were at the place of meeting.   In the reception room Sir Henry met Mr. J. Nasmyth, of steam hammer celebrity, and showed him a specimen of the metal which his process produced.   Mr. Nasmyth was delighted to see it and said : " You will reap a rich reward for this, and you thoroughly deserve it."   In reference to the specimen of metal shown him, Mr. Nasmyth said : " That is a real

British nugget "—he was so struck with its excellence. Sir Henry's paper was taken first that morning in the Mechanical Section, of which Mr. G. Rennie was the president. After some pre-liminary remarks on the extent and objects of his experiments, he said : " On this new field of inquiry I set out with the assumption that crude iron contains about five per cent. of car-bon ; that carbon cannot exist at a white heat in the presence of oxygen without uniting therewith and producing combustion ; that such combustion would proceed with a rapidity dependent on the amount of surface of carbon exposed ; and, lastly, that the temperature which the metal would require would be also dependent on the rapidity with which the oxygen and carbon were made to combine ; and, consequently, that it was only necessary to bring the oxygen and carbon together in such a manner that a vast surface should be exposed to their mutual action in order to produce a temperature hitherto unattainable in our largest furnaces." He then proceeded to give a lengthy account of the way in which this was carried out in the converter ; adding a good many details in the process that he had worked out between making the first experiment we have described and the reading of this paper. He next described the properties of the metal thus produced, and the most important purposes for which it could be used ; and in conclusion stated that its cost would not be much different from that of ordinary iron.

Such was the first public announcement of the Bessemer process made at Cheltenham on August 11th, 1856. As soon as the paper was read Mr. James Nasmyth rose and expressed his approval of the principles of the invention. Next rose the gentleman who had that morning ridiculed the idea in the inventor's presence at the breakfast table. He now stated that he was already so impressed with the prospects of the invention that he would place the whole resources of his large iron making establishment at the inventor's disposal at once in order to work the process on a large scale. To others in the trade it appeared too good to be true. "It is difficult," observed Mr. Isaac Lowthian Bell nearly twenty years afterwards, "to say whether science or art was more perplexed at the announcement of the Bessemer process. The former appears to have thought it prudent to remain silent, at all events in the Transactions of the British Association—for all the notice there bestowed on the discovery is the bare mention of the title of his communication. Art was less reticent, for I remember the ridicule with which the proposal was received."

The paper of Sir Henry Bessemer was read on Monday morning, and on the Thursday following it was published in the *Times*. At the Dowlais iron works, then the largest in the world, it excited so much scepticism and curiosity that the heads of departments there determined instantly to test its— uselessness! A vessel, answering the purposes of a converter, was fitted up next day, and filled with fluid metal, while cold air at high pressure was injected in the manner described in the paper. Their doubts were soon dispelled. It worked amazingly well, and before that week had ended two bars, twenty-five feet long, were rolled in the mills.

Baxter House, St. Pancras, was visited by many inquirers anxious to see this wonderful invention in operation. Eight days after the report of Sir Henry's paper appeared in the *Times* the process was again tested in the experimental converter in the presence of several leading iron-masters, practical engineers, and scientific men. Nearly 7 cwt. of molten iron having been poured into the converter, it soon began to boil, while air—the food of fire—was blown through it. The object, according to the account given at the time, was to produce a mass of cast steel rather than to continue the process to the extent necessary for making pure iron free from carbon; and, therefore, "the blow" was only continued for twenty-four minutes. Two small specimen ingots were first drawn off, and then the remaining mass of metal was run into a mould in the floor, forming an ingot of nearly 6 cwt. The trial was pronounced a great success ; yet some of the spectators thought Sir Henry was too enthusiastic when he told them that the semi-steel produced by his process would probably in time supersede malleable iron for railway purposes, and that the process of forging and welding, which, in ordinary practice, was necessary whenever a piece of iron work of a larger size than 80 lbs. or 100 lbs. was required, would be dispensed with. However, everything for the present looked promising. The Emperor Napoleon, who had been his greatest patron at the outset, on hearing of the marvellous success of the process intimated his intention of bringing it into practical operation in the arsenal at Roulle ; and at Woolwich, where the authorities at the first mention of it cynically alleged that Mr. Nasmyth had previously made the same discovery, a disclaimer from that gentleman was accepted, and Sir Henry's originality was undisputed.

On the 1st of September another trial of the process took place at Baxter House, when the following incident, narrated by one who was present, occurred : One amongst the gazers from the iron districts—a stout, wealthy-looking, growling individual, with a spice of St. Thomas in him—thought the casting too hot to try with his fingers ; but expressed his belief that it was not malleable but simply cast iron. On this Sir Henry Bessemer spoke not, but entering the shed returned with a large axe, thick on the edge, wherewith he " laid on load." Two cuts at the edge of the ingot left an impression in indents analogous to those produced in chopping a wooden post. " That's not cast iron," growled some one, " such as thou wisheth it had been ; " and the Staffordshire chieftain, Mr. Blackwell, possessing himself of a piece, subjected it to cold hammering on the anvil, and subsequently to filing in a vice—the file hanging to it as to tough copper. " We must change our proceedings," were the words that followed ; " whether by this process or some other not yet known, it is clear that we cannot go on as we have done."

Experiments tried elsewhere had very different results, and gave rise to very damaging reports. For instance, in the first week in September some experiments made by Mr. Clay at the Mersey Steel and Iron Works were an undoubted failure. Mr. Clay reported that the iron produced was rotten hot and rotten cold—in other words, useless. Detractors and unbelievers of course made the most of these failures.

In the conflict of opinions that was raging as to the success or failure of the new process one eye-witness of its success gave an account of it so prescient and pregnant that it has now a historic interest. Dr. R. H. Collyer, writing on September 10th (1856) from Park Road, Regent's Park, said :—

"I availed myself last Friday of the public invitation of Messrs. Bessemer and Longsden to visit Baxter House and inspect their new process in operation. I found there assembled some seventy or eighty of the most eminent persons connected with the manufacture of iron—that metal which, in point of importance to the wants of civilised men, causes all other metals to sink into comparative insignificance ; the great onward progress of the age—our railroads, our steam-engines, and a thousand appliances of human ingenuity—would all have been in a state of inertion had not the use of iron been so plentifully vouchsafed for man's necessities. Any improvement, therefore, must be held as a triumph of human progress, a step towards that period when the toil, the wear and tear of muscle will cease. Thousands of the race who now are rendered decrepit in their youth will, by the employment of this metal, in countless and inconceivable forms, increase the longevity of the species. Man's life now is concentrated, as it were, into centuries—time is merely conventional ; that which only a quarter of a century since took us weeks to effect is now the work of days. The extended use of this, the most important of all the metals on the earth's surface—iron— will cause a man of 50 to have actually lived to the age of 150 in comparison with the standard of his ancestors. Mr. Bessemer has undoubtedly achieved, by the application of a known, simple, and beautiful idea, one of the greatest triumphs of the age. All metals, according to their purity, become ductile and malleable," &c. After describing the details of the operation, he goes on to say that after a twenty-four minutes' blow in the converter, " the blast was turned off, and the purified metal run into an ingot. The whole experiment was conducted with a degree of precision and neatness which would have done honour to a Faraday, a Turner, or a Dalton. I must not forget to mention that a bar of malleable iron was exhibited. It was about twelve feet long, half an inch thick by two inches wide. This was bent and showed great ductility, though, on fracture, as the ironmasters said, the grain exhibited was not such as would lead one to expect such ductility. It seemed to them a paradox. The fact is the great secret is getting metallic purity. The particles don't require to be interlaced or fibrous to the same extent as when iron contains even a small proportion of phosphorus or sulphur. The former I consider the most pernicious of all. I would suggest, with due deference, that a stream of finely pulverised anhydrate of lime (dry lime) be forced at a given time with the compressed air into the incandescent mass of iron. The lime having a great affinity for silica (sand) and phosphorus would form a phosphate and silicate of lime, and be thrown off with the slag. By this contrivance I cannot conceive but that the phosphorus would entirely be got rid of . . . I do not believe the process is complete, but chemically no one can doubt the great move made in the right direction. In conclusion allow me to state that I cannot but pity the host of petty detractions, jealous rivalries, and difficulties which suggest themselves to that class of men who never see anything till every one else laughs at their intellectual blindness."

Dr Collyer's gratuitous suggestion for perfecting the chemical purity of the Bessemer process lay dormant for nearly a quarter of a century; and its practical application then by one who, in 1856, was a child six years old was considered an achievement only surpassed by the parent invention of Sir Henry Bessemer. It added millions sterling to the world's wealth.

To resume, on September 14th one of the bars made at Woolwich from Bessemer iron was heated and rolled out into a thin sheet at one of the Wolverhampton mills. It was then taken to the workshop of a tobacco-box maker, who punched it cold into the required shape for a tobacco-box. To the surprise of the operator it worked perfectly. The box was completed, and on applying the polishing tool a polish was produced that the operator described as equal to that of steel. "A better bit of iron," he said, " I never worked."

Meanwhile numerous overtures were being made to the success· ful inventor for permission to work the process in this country. The managers of the Ebbw Vale Iron Works, who were on the point of starting new works for the production of steel by another process, offered £50,000 for the patent, but this was declined. Sir Henry determined to work the process in another way. He divided Great Britain into five principal iron districts, and announced that he wanted one ironmaster in each district to have so great an interest in the successful result of his invention that he would always act for him instead of against him. He proposed that any ironmaster who was the first to apply in his district for a license should, by paying one year's royalty on a quantity to be decided by himself, pay no other royalty during the fourteen years of the patent; so that he would be interested very strongly in maintaining the patent, improving it, and making it a nucleus of operation in his district. This proposal was accepted by five different ironmasters; two of them paid £10,000 each; and the licenses sold within three weeks of the reading of his paper at Cheltenham amounted to £25,000.

The Dowlais Iron Company were the first to take a license; and it was arranged that Sir Henry should advise them as to the details of working the process. Mr. Menelaus, then the manager of the works, in his first interview on the subject said to the inventor: "We have seventeen furnaces in blast, and I will tell you the burden of each of them. You select your furnace, and if it is possible to put up your apparatus before it we shall do so." Sir Henry replied: "It does not matter where you put up my apparatus; it will work any kind of iron." At that time there happened to be the remains of a cast house standing before one of the furnaces, and as it was easy to roof over the space in front of that furnace cheaply and quickly it was agreed to put up the converter there. The furnace was at the time making iron for common rails. This iron, in its fluid state, was then run direct from the furnace into the converter, where it blazed, sparkled, bubbled, and showed all the beautiful phenomena of the process. The whole operation looked very satisfactory; but when they came to work the metal produced they were surprised to find that it was utterly useless for any purpose. This appeared inexplicable, so the experiments were repeated, and, although a good deal of money was spent in this way, the success of the first rude experiment, made a day or two after the announcement of the process, was never equalled. Sir Henry consequently left Dowlais with serious apprehensions as to the success of his invention.

The bright prospect which the first announcement of the process raised was now overcast, and was eventually followed by a general gloom. An invention which was at first received with a shout of triumph was, two months afterwards, declared to be impracticable. One journal after another ran it down, and so general became the chorus of denunciation that the inventor himself filled a portfolio with cuttings from scientific and industrial papers written to demonstrate that his process was the work of a visionary.   Worst of all was the mysterious failure of the process under his own direction.   Within a few weeks after the first account of it was given at the British Association, experiments were made by several ironmasters who turned a low pressure blast into a basin roughly adapted for the purpose.   Singularly enough every such attempt failed.   No one knew why.   The experimenters thought they followed the directions of the inventor; but they did not succeed.   Their high-flown expectations were disappointed, and a revulsion of feeling naturally followed.   Exactly six weeks after the first description of it appeared in the *Times* a meeting of the ironmasters of South Staffordshire and East Worcestershire held at Dudley—then the centre of the iron trade—condemned the new process as a practical failure.

CHAPTER III.

" If geometry contained the rule of life, there would be men found to dispute its axioms."—LEIBNITZ.

"IT is a truly strange coincidence," said David Mushet, then considered an authority on questions connected with iron and steel, "that this magnificent invention (the Bessemer process) should be brought before the world at the very moment that we have been urging the nation to present some grateful recompense to the family of the father of the founder of the British iron trade. I do not hesitate to express my conviction that Mr. Bessemer has produced the grandest operation ever devised in metallurgy. I trust the discoverer will reap his due reward. . . . It is a painful reproach to this country, that most of its great inventors in iron and steel have been recompensed with ruin. Envy and rivalry have been aided by the labyrinthical decisions of our courts of law in patent cases. After struggling for fourteen years through all their tortuosities, patentees have arrived at last in the supreme court of appeal merely to see their hopes finally extinguished, and in the face of a majority of judges aiding the deliberations the decisions of four previous courts reversed, in order that one peer might reassert the opinion he had maintained as a commoner some dozen years before. As late as 1854 such a case occurred. I sincerely trust the flowing sail of Mr. Bessemer's success will drift him upon no such rocks and quicksands."

E

These words were first published on the 22nd of August, 1856. Let us look at this strange coincidence. At that time a series of letters and articles were appearing in the *Times*, describing the miserable condition of the family of "the author of those improvements in the manufacture of iron to which Great Britain owes her position among the nations." It was the inventions of Henry Cort, for refining and rolling iron, that made this country independent of Russia and Sweden for supplies of iron. In the contemporary literature of that day it was stated, by one who spoke with authority, that the year 1785 brought before the world the *Times* and Henry Cort's inventions, and that they had acted together for seventy-one years as the two greatest engines of modern civilisation ; while Mr. G. R. Porter stated that since 1790, when Cort's improvements were completely established, the value of the landed property of England had doubled. Now, in 1856, the public and Parliament were urgently entreated to consider the fate of this national benefactor, who had received no direct benefit from his inventions. Eight days after the first announcement of the Bessemer process, David Mushet, in calling attention to the case of Cort's family, said the obligations under which the British iron trade lay to Henry Cort were notorious. " Forty-one firms, in 1811, subscribed to a resolution thanking the deceased for his two great inventions, and contributed a sum in relief of the distress of his widow and nine orphan children. It is true, the contribution amounted to only one-twentieth part of a farthing in the pound upon the profits these inventions had then realised to them ; but it was at least a fair and open acknowledgment of the obligation and the debt. The profits which these manufacturers have since that date realised by puddling and rolling have been ten times greater than in the previous twenty-five years, and cannot fall much short of 30,000,000*l.* sterling. A second contribution of one-twentieth part of a farthing in the pound on these profits would realise an ample sum for the family of their benefactor; but a subscription of only the same sum as that of 1811, which would be at the rate of one two-hundredth part of a farthing in the pound on the whole profits would alleviate the immediate distress of the inventor's aged and only surviving son, and give him time and ease to bring properly before the legislature in the next session his unparalleled claims upon the credit of the nation at large." One ironmaster, to whom an appeal was made, replied that "It is truly a sad case; but it would happen every day in the year were there as many Corts as days."

Such was the account given of the recompense accorded to the labours of the man whose inventions were the mainspring of the malleable iron trade at the time when Sir Henry Bessemer was trying to perfect an invention that would enable steel to supersede malleable iron. For the present the prospects of the one seemed as dismal as those of the other had ever been; and the way in which the fate of Cort was now being paraded before the world was anything but encouraging to an inventor who was seeking to lead his country a great step further forward. Nor was this all; during the six weeks that followed the announcement of the Bessemer process one inventor after another called attention to the neglect with which an ungrateful public had treated *his* meritorious invention for the improvement of the steel trade. These claims came from America, France, and Austria, as well as England. Public opinion, which professes to be so superior to any taint of partiality that it generally represents its image of justice as born blind, now added the Bessemer process to the list of visionary schemes whose proper end was oblivion. Against this verdict one man dissented—the undaunted inventor. He saw that there was a difficulty in the working of his process, and could not find out the cause of it. Still he had faith in it; and instead of answering the attacks made upon it in the press, he determined to investigate the cause of this failure. Adequate knowledge, says George Eliot, will show every anomaly to be a natural sequence. Acting on this principle Sir Henry, aided by the additional capital he had obtained through the sale of licenses, quietly set to work to investigate the problem upon the solution of which not only his hopes but his fortune were staked, and he did so not by mere laboratory experiments, but by the actual working of from half a ton to two tons at a time.

Meanwhile the patient inventor was not the only one who studied the question. Before the year 1856 had closed, a writer in a Birmingham journal said, with reference to the new process : " It is specially important that accurate chemical analysis should be resorted to in order to show the composition of this iron, and to show that the new process will truly purge it of sulphur and phosphorus—elements the presence of one per cent. of which is fatal to the quality of the iron. So far as we are aware, this important information has not been communicated to the public ; and so long a time has now elapsed that we despair of receiving it from the quarter it was most naturally expected from. In the hope of contributing to the settlement of a question which has already so long disturbed the public mind, we have imposed upon ourselves a task which we think should have been spared us, and present to our readers such an analysis of Mr. Bessemer's iron as we have been daily hoping to see published by that gentleman himself. The specimens we have experimented upon possess those physical properties which, from repeated descriptions, the public are sufficiently familiar with. The iron consists of an agglutinated mass of large brilliant crystalline grains, possessed of a very imperfect malleability, flattening under the blow of a hammer, but almost invariably cracking at the edges. It is wholly destitute of a fibrous structure, and only after having been repeatedly heated and drawn out in a smith's forge, exhibits the properties of an inferior wrought iron. In contrasting the change effected by Mr. Bessemer's treatment with that of the refinery, the following particulars force themselves strongly upon our notice. Mr. Bessemer's method removes most effectually the carbon and silicon, while in the refinery these are but little diminished. The carbon is eliminated with a perfection that we should scarcely have thought possible, but we are without information as to the sacrifice at which it has been effected; the amount of iron oxidised by the vivid combustion which Mr. Bessemer induces we are unable to ascertain. The point which most prominently strikes the chemist in Mr. Bessemer's iron is the large amount of phosphorus which it contains—an amount utterly fatal, we fear, to the value of Mr. Bessemer's method. His treatment, we suspect, does not sensibly diminish the amount of this element; but this, too, is a point on which we must be dependent on Mr. Bessemer. It is by the puddling process that the phosphorus and sulphur are mainly removed. As yet, so far as we can learn, Mr. Bessemer has done nothing towards the removal of this pernicious element—phosphorus; and in this important respect his process must be regarded as a failure."

These observations were in the main correct; but in the shoals of condemnation then poured upon the process one might be said to wade through "a continent of mud" before coming to this little bit of solid ground. Moreover, this criticism, however correct, concluded in a way that was scarcely likely to commend it to the attention of Sir Henry Bessemer. " In taking leave of this subject," said the critic, " we think we may safely predict that the iron manufacture will remain unaffected in any essential respect by anything which Mr. Bessemer has done. No one will more sincerely rejoice at any real improvement in the iron manufacture than we shall, although we admit a preference for such improvements as are not heralded by announcements of ' revolution,' but are modestly propounded and left to demonstrate their importance by that quiet and cautious induction into practice which generally characterises really meritorious inventions. We fear, however, that Mr. Bessemer has lost his opportunity. The interest he has wasted cannot readily be reawakened, and less readily by him than by any other. Like the shepherd in the fable who cried 'wolf, wolf,' when there was no wolf, and whose cries for help were unheeded when the marauder indeed came, Mr. Bessemer will find that any real improvement he may hereafter make will suffer a neglect for which his own hastiness and want of caution will be alone to blame."

Sir Henry, while inundated with candid advice of this sort, continued to investigate everything for himself regardless of all suggestions. The two extracts that we have given show that some ideas of permanent value were freely offered to him but were set at naught. It was not till another series of independent experiments were made that he himself discovered the secret of failure. It then appeared that by mere chance the iron used in his first experiments was Blænavon pig,[1] which was exceptionally free from phosphorus ; and consequently when other sorts of iron were thrown at random into the converter the phosphorus manifested its refractory nature in the unworkable character of the metal produced. Analyses made by Professor Abel for Sir Henry showed that this was the real cause of failure. Once convinced of this fact, Sir Henry set to work for the purpose of removing this hostile element. He saw how phosphorus was removed in the puddling furnace, and he now tried to do the same thing in his converter. Another series of costly and laborious experiments were conducted ; and first one patent and then another were taken out, tried, and abandoned. His last idea was to make a vessel in which the converting process did not take place, but into which he could put the pig iron directly it was melted, along with the same kind of materials that

---

[1] In the process of smelting, the principal channel along which the metal in a state of fusion runs when let out of the furnace is called the *sow,* and the lateral channels or moulds are denominated *pigs;* whence the iron in this state is called *pig-iron* or cast-iron.

were used in the puddling furnace. He was then of opinion that he must come as near to puddling as possible, in order to get the phosphorus out of the iron. Just as he was preparing to put this plan into operation there arrived in England some pig iron which he had ordered from Sweden some months previously. When this iron, which was free from phosphorus, was put into the converter, it yielded in the very first experiment a metal of so high a quality that he at once abandoned his efforts to dephosphorise ordinary iron. The Sheffield manufacturers were then selling steel at 60*l*. a ton, and he thought that as he could buy Swedish pig iron at 7*l*. a ton, and by blowing it a few minutes in the converter could make it into what was being sold at such a high price, the problem was solved.

Sir Henry Bessemer has preserved a remarkable proof of the success which he now attained. It is a small cannon ; and though only weighing two and a half hundredweight, it was the first gun ever made of malleable iron without weld or joint. Nor is this its only distinction. An analysis of it made by an eminent metallurgical chemist showed that it contained 99.84 per cent. of pure iron. This is almost perfection. According to analyses given by Dr. Percy, the famous Dannemora Swedish iron, which had the reputation of being the finest quality in the world, contained only 99.312 per cent. of pure iron. In other words, the six impurities that impregnate iron only constituted one part in 625 of this specimen of Bessemer iron, while in the Dannemora iron they constituted one in 145 parts.

But there was yet one thing wanting. He had now succeeded in producing the purest malleable iron ever made, and that, too, by a quicker and less expensive process than was ever known before. But what he wanted was to make steel. The former is iron in its greatest possible purity ; the latter is pure iron containing a small percentage of carbon to harden it. There has been an almost endless controversy in trying to make a definition that will fix the dividing line that separates the one metal from the other.

Sir Henry Bessemer has well observed that from a chemical point of view the line of demarcation which separates these substances is as little marked as the rainbow's hues, which melt imperceptibly into each other, leaving no part at which it may be said here one ceases and there the other begins: Thus it is with iron and steel, which passes by almost imperceptible gradations from grey iron, through every stage of mottled and white, to hard steel, and from this to steel in its mildest form, which passes into malleable iron almost unmarked.

For our present purpose suffice it to quote the account given in a popular treatise on metallurgy published at the time when Sir Henry was in the midst of his experiments. " Wrought iron," it says, " or soft iron may contain no carbon ; and, if perfectly pure, would contain none, nor indeed any other impurity ; this is a state to be desired and aimed at, but it has never yet been perfectly attained in practice. The best, as well as the commonest foreign irons always contain more or less carbon. . . . Carbon may exist in iron in the ratio of 65 parts to 10,000 without assuming the properties of steel. If the proportion be greater than that, and anywhere between the limits of 65 parts of carbon to 10,000 parts of iron and 2 parts of carbon to 100 of iron, the alloy assumes the properties of steel. In cast iron the carbon exceeds 2 per cent., but in appearance and properties it differs widely from the hardest steel. These properties, although we quote them, are somewhat doubtful ; and the chemical constitution of these three substances may, perhaps, be regarded as still undetermined." Now in the Bessemer converter, the carbon was almost entirely consumed. In the small gun just described there were only 14 parts of carbon for 1,000,000 parts of iron. Sir Henry's next difficulty was to carburise his pure iron, and thus make it into steel. " The wrought iron," says Mr. I. L. Bell, " as well as the steel made according to Sir Henry Bessemer's original plan, though a purer specimen of metal never was heard of except in the laboratory, were simply worthless. In this difficulty a ray of scientific truth, brought to light one hundred years before, came to the rescue. Bergmann was one of the earliest philosophers who discarded all theory, and introduced into chemistry that process of analysis which is the indispensable antecedent of scientific system. This Swedish experimenter had ascertained the existence of manganese in the iron of that country, and connected its presence with suitability for steel purposes." Manganese is a kind of iron exceptionally rich in carbon, and also exceptionally free from other impurities. Berzelius, Rinman, Karsten, Berthier, and other metallurgists had before now discussed its effect when combined with ordinary iron; and the French were so well aware that ferro-manganese ores were superior for steel-making purposes that they gave them the name of *mines d'acier.*

In 1830 a patent was taken out for the application of manganese in steel making by Mr. Josiah Marshall Heath, whose adventures in connection with this subject are worthy of record here. He was a talented man in the service of the East India Company, and when he went to his post in India he was so well versed in Oriental literature that a Sanscrit professorship was offered to him before he had attained the age of twenty-one. A friend having requested him to procure some steel heads for boar spears, he was thus led to visit the steel works in India, and was so much struck with the clumsiness of the process of steel making that he determined to study the subject. In traversing the Malabar coast he discovered great masses of iron ore, which he thought could be converted into the cheapest steel for the supply of Europe. But, on inquiry, he found that experienced metallurgists pronounced them unmanageable, it being said that these ores could not be converted into steel or iron. Heath thought otherwise; and not only resigned his appointment, but devoted his private fortune and pension to the investigation of this matter.   He visited all the great iron and steel works in the world, and thus acquired a practical knowledge of the trade.   As the result of his experiments and researches he took out his patent, which he described as an invention for the use of carburet of manganese in any process for the conversion of iron into cast steel.   His fortune was now spent, and though his resources were exhausted, his invention did not make any progress between 1830 and 1840. In the latter year he visited Sheffield, where he succeeded in introducing it, and where it was soon reported to have made a valuable improvement in the quality of the steel produced.   He, however, confided the secret of his success to his agent, telling him that a small percentage of oxide of manganese with a little coal tar put into the crucible in which the steel was to be fused would have the same effect as his patent carburet of manganese. Not long after this disclosure he discovered that his agent had established steel works, and was making steel by the application of the information confidentially communicated to him.   All claims for remuneration on account of infringement of patent rights were ruthlessly disregarded.   The case was carried into the law courts by the luckless patentee, who in the first trial of 1843 was non-suited.   It was again tried in 1844, when the Court of Exchequer gave a verdict for Heath on all the issues.   But this verdict, after three years' consideration, was set aside, and a new action raised in the Court of Common Pleas, which was not tried till 1850.   When at last the day of judgment came in the Common Pleas, Mr. Justice Cresswell, "mindful," says Mr. Dickens, "of the etiquette of the bench, declared that he could not, sitting singly, confirm or reverse the judgment of the Court of Exchequer; but that he would direct the jury to find for the defendant, and the plaintiff would then be at liberty to bring the whole matter again before a competent tribunal.   Mr. Heath procured a stall at the Great Exhibition of 1851, and arranged with his own hands his rare metallurgic specimens; but before the Exhibition opened, and before his case came again to be argued, his weary heart ceased beating. He died! leaving his successors to prosecute the claims they derive from him."

The use of manganese in steel making then ceased to be a "burning question" until the announcement of the Bessemer process in 1856 revived it in another form. Sir Henry Bessemer discovered that by introducing a small quantity of ferro-manganese into his converter, sufficient carbon was imparted to the metal to give it the properties of steel of a mild quality. The circumstances that led up to this discovery, as narrated by himself, form not the least interesting episode in the history of his invention. In acquainting himself, in the early stages of his experiments, with the methods then in use for manufacturing iron and steel, he read an account of Heath's method of using manganese and its introduction into England. He learned that after many experiments Heath had succeeded in producing metallic manganese on a commercial scale, but at that time the properties of that valuable metal were little understood except by Heath. The metal in a granular state (a sort of physic, as it was called) was sold in packages by Heath, who instructed the steel makers of Sheffield to put a certain weight of it wrapped in a little piece of paper into each crucible during the operation of melting the steel. It was found that very inferior brands of iron, compared with those generally used in England for making cast steel, could be made into excellent steel by the use of this metallic manganese. It cheapened the production, while it did not injure the quality of the material produced; on the contrary, it conferred on it the property of welding, and increased the malleability of the metal. Heath sold the metallic manganese which he produced in his own establishment at a price which was very much greater than it cost when made in the steel crucible itself; and when, therefore, the Sheffield manufacturers were informed that the same result could be produced in the way Heath told his perfidious agent, they said this simpler method of using it was not included in his patent, and they consequently ceased to pay him any royalty. Having become acquainted with these facts, Sir Henry Bessemer says he suspected—though he did not know it thoroughly in the early stages of his process—that manganese was a material that would materially help him also, because, like Heath, he was endeavouring to introduce an inferior brand of iron and make it into cast steel of a mild and malleable quality. Seeing that he could not introduce manganese and carbon separately into his converting process, he commenced some experiments on the reduction of ordinary manganese to the metallic state. In doing so he found considerable difficulty, which chiefly arose from the fact that he had powdered the charcoal and mixed it with the powdered hematite iron ore and the manganese. These substances, being in the condition of powder, did not allow the minute particles as they were reduced to the metallic state to fuse and run into the mass, but he subsequently found that by putting in the materials in moderate sized lumps—the size of peas or larger—excluding all the dust, and adding little lumps of charcoal, he could gradually reduce the hematite ore and manganese in the crucible to the metallic state. They ran readily through the lumps of charcoal to the bottom of the crucible ; and thus was produced ferro-manganese. " Just about that period," says Sir Henry, " a patent was taken out by Mr. Robert Mushet, for the introduction of manganese into my process, precisely the same as that which was in daily use in Sheffield—that is, he mechanically mixed the powdered manganese with pitch, and then stamped the mixed material to a powder, which, he proposed, should be blown in at the bottom of my converting vessel through the tuyeres or air pipes. In these conditions no reduction of the manganese would take place ; and such a mode of proceeding was utterly valueless." However, a further patent was taken out by Mr. Mushet for what he called a triple compound ; it was for uniting carbon, iron, and manganese. This patent barred the way to further experiment for a time ; but on investigation Sir Henry discovered that a triple compound of iron, carbon, and manganese had existed for many years in almost innumerable samples of pig-iron, and in various proportions, especially in the metal called spiegeleisen, used so largely for steel making in Prussia, where the manganese amounted to something like seven or eight per cent., with four per cent. of carbon and eighty-eight of iron.

Meanwhile, Mr. Mushet took out several patents that were intended to cover every possible mode of putting that metal into steel made in the Bessemer converter ; and shortly after Sir Henry began to use manganese, Mr. Mushet's partner requested him to take out a license for its use. " In reply," says Sir Henry, " I read to him Heath's patent and part of my own specification, in which I mentioned the use not only of manganese, but of many other substances ; and I informed him at the same time that I did not believe his claim could be sustained. I also offered, if he would come to Sheffield with his solicitor, and would bring two witnesses, that I would make the steel by my process with the assistance of manganese in their presence, that I would cast it into ingots, that I would take it into the town, sell it in the open market, and give him the invoice, so that there could be no question as to my having infringed his patent if it was a valid one. That offer, however, was not accepted, and I was never afterwards warned or even desired not to use Mushet's patent." By the time the Bessemer process had got into practical operation its prospects had sunk so low in public estimation that it was not thought worth while paying the £50 stamp due at the expiration of three years on Mr. Mushet's batch of manganese patents, which were consequently allowed to lapse.

Mr. Mushet's account of the manganese incident is, of course, different from that of Sir Henry Bessemer. It is, however,

interesting.   He says : "When Bessemer read his celebrated
paper at the Meeting of the British Association at Cheltenham,
in 1856, I saw clearly where his difficulties would arise, and that
he could not by his process produce either iron or steel of commercial value.   A few days after the reading of the paper I
received specimens of Bessemer metal.   Some of these were
cold short (brittle when cold), and some were cold tough, but all
were alike red short (brittle, or breaking short when red hot) at
any heat under the welding heat.   They were ductile enough when
worked at a high welding heat ; but as soon as the temperature
was lower the bars broke off or crumbled like heated cast-iron.
I at once saw that by melting them again with manganesic pig-
iron or spiegeleisen they would form good steel or iron, according
to the dose of manganesic pig added to them.   Late that night
it occurred to me that by mixing the already melted Bessemer
metal with melted spiegeleisen the process could at once and
simply be rendered successful.

"I immediately lighted a fire in a small steel melting furnace,
and placed in the furnace two crucibles, one containing eight
ounces of Bessemer metal, and the other one ounce of spiegeleisen.
When the contents of the crucibles were melted I withdrew the
crucibles, and poured the melted spiegeleisen into the melted
Bessemer metal ; I then emptied the mixture into a small ingot
mould.   The ingot was piped, and had all the characteristics of
an ingot of excellent cast steel.   I next heated the ingot to a
fair cast-steel heat.   Mrs. Mushet held it in a pair of tongs, and
I drew it out with a sledge hammer into a flat bar.   I heated
this bar, and then twisted it in a vice, at a white heat, red heat,
and black-red heat ; and it remained perfectly clear and sound in
the edges, without a trace of red shortness.   I now doubled and
welded the bar, and forged it into a chisel, which I tempered
and tried severely, for a flat chisel and diamond point, upon hard
cast-iron.   The chisel stood the test well, and was in fact welding

cast-steel worth at the least 42*s.* per cwt. I saw now that I had made a discovery even more valuable than that of the Bessemer process ; for, although the Bessemer process was not of any value apart from my invention, on the other hand my invention could be applied extensively in the manufacture of pot-melted steel. Less conversant with the world than with matters relating to iron and steel, I confided in certain parties of great wealth and influence in the iron trade, believing that I had to deal with men of honour and integrity, incapable of a mean and base action. On this score I gained experience, which cost me my patents, and taught me a lesson not easily forgotten. I placed my patents in the hands of parties who promised to carry them out, and see justice done to me. I now proceeded to extend the scale of operations as follows :

" I charged sixteen melting-pots with 44 lbs. each of Bessemer metal, and when this was melted I poured into each pot 3 lbs. of melted spiegeleisen. I then poured the contents of the sixteen melting-pots into a large ingot-mould, and the ingot thus made was sent to the Ebbw Vale Ironworks, and then rolled at one heat into a double-headed rail. The rail was sent to me for inspection and to be by me forwarded to Mr. Ellis at Derby station, to be laid down there at a place where iron rails had to be changed once in three months. When the rail, which was very perfect, came to me, it was so thickly studded with the words, ' Ebbw Vale Iron Company,' that no space remained to squeeze in the words, ' Robert Mushet.' I sent the rail to Derby, and I have read a public statement made by Mr. Adams of the Ebbw Vale Ironworks to the effect that this rail remained as perfect as ever after six years' wear and tear from the passage over it of 700 trains daily. That was the first Bessemer rail.

" I next charged twenty melting-pots with 46 lbs. each of Bessemer metal, and when melted I poured into each pot 2 lbs. of melted spiegeleisen, and then emptied the contents of all the melting-pots into one ingot mould. The ingot was rolled at two heats at the Ebbw Vale Ironworks into a bridge rail, which would have been about 30 feet long ; but the engine was over-powered when the rail was in the last groove, and stopped, so that the rail had to be cut in two. One piece, 16 feet long, was exhibited in the office of an influential iron company in London as the produce of the Uchatius or atomic process of steel melting. Let us charitably suppose that the gentlemen who exhibited the rail were at the time labouring under some mental hallucination. To enable me to specify my patent, I was very generously furnished by Mr. S. H. Blackwell, of Dudley, with a blowing machine capable of sustaining a pillar of blast of 10 lbs. per square inch. With this blowing apparatus and some small furnaces operating upon from 60 lbs. to 600 lbs. of melted cast-iron, I experimented for six months, and satisfied myself that the whole affair was as simple, and indeed far more simple, than the ordinary foundry process for melting and casting iron."

As to Heath's use of manganese, Mr. Mushet says : "Mr. Heath was one of my father's most intimate friends, and always consulted the latter on metallurgical points, though he seldom followed his advice, being somewhat opinionated. I have Mr. Heath's correspondence with my father, and with myself latterly, and Mr. Heath states in his letters that he took the idea of his manganese patent from the experiments on oxide of manganese and iron made by my father, the details of which experiments appeared first in the *Philosophical Magazine*, and subsequently in my father's papers on 'Iron and Steel.' Mr. Heath laid his patent process before my father, and asked his advice. That advice was to patent the use of oxide of manganese, but, unfortunately for himself, Mr. Heath did not adopt this advice, and his patent was lost in consequence."

Never, perhaps, were two rival inventors more confident in the justice of their respective "rights." Mr. Robert Mushet never ceased to proclaim that he was the first to apply manganese to Bessemer metal; and Sir Henry Bessemer was never proved to have infringed any patent right by the free use of manganese. But the one inventor reaped a rich reward for his labours, while the other earned nought. The opinion of metallurgists in later years was that both had worked successfully at the same problem; and as some compensation for the disparity of fortune that attended the labours of the disappointed chemist, Sir Henry Bessemer generously presented him with an annuity of £300 a year:

" Gentler Knight there never drew a lance."

The history of inventions in relation to the use of manganese did not end with its application in the Bessemer converter. It might rather be said to have only begun there. When an interval of some time had elapsed after Sir Henry Bessemer discovered the use and manufacture of ferro-manganese, he began to see the necessity of manufacturing that material on a large scale for his process, because, as he said, it would enable him to make a very mild steel, and because the use of spiegeleisen would not do so owing to the large quantity of carbon in it. At that time he heard that large quantities of manganese were being used in Glasgow in the St. Rollox Chemical Works; and he went to Glasgow to ascertain whether the proprietors of those works would undertake to manufacture ferro-manganese for him with their waste manganese. A friend whom he called upon with the view of obtaining an introduction to the proprietors in question brought the subject under the notice of an able chemist—Mr. Henderson—to whom Sir Henry Bessemer explained what he required. Mr. Henderson undertook to produce the material wanted, and Sir Henry accordingly left the matter in his hands. Mr. Henderson did not at first agree with Sir Henry's view of the problem—that if he got manganese, or ferro-manganese rich in manganese, the carbon would be less. He therefore first

F

investigated this point, because unless they got the carbon reduced it would be of no great use in making soft steel.   A few crucible experiments confirmed him as to the correctness of Sir Henry Bessemer's view.   He found that as the manganese increased the carbon decreased; but when he attempted to manufacture ferro-manganese on a large scale he found that it was extremely difficult.   The intense heat required and the excessive affinity of oxide of manganese for bricks were so great that for a long time the problem baffled him.   He first tried one refractory substance and then another for the bottom of his furnace, but they were all melted away.   Repeated failures brought him to the verge of despair.   In this dilemma he happened one day to mention his difficulties to another chemical manufacturer, who suggested that he should use bricks made of carbon.   He at once determined to test that idea.   Accordingly, he took some hard coke and ground it very fine; he then mixed it with just enough of tar to cement the particles together; and having put it in an iron mould he heated it to a red heat, and thus obtained fine solid carbon blocks.   These were built into a furnace, with which he had henceforth no difficulty.   The first bottom of that description was worked for three years, and even then on removal it was found to be scarcely worn an inch.   No sooner, however, had he succeeded in the manufacture than the works in which the furnace was built came to grief.   The Tharsis Company bought the patent from him, but as they did nothing with it, he afterwards explained his plan at the Terre Noire works in France, where the manufacture of ferro-manganese was for some years exclusively carried on by a more perfect process, which raised the yield of manganese from 25 to 75 per cent., and reduced the price 50 per cent.   It is now manufactured at several large English iron works.

To show the superior properties of the mild or soft steel produced by the introduction of a small quantity of manganese

when the liquid metal had been purified by the intense combustion caused by the cold blast, Mons. Gautier, of Paris, gives
the following comparative statement of the average resistance
of steel manufactured in the Bessemer converter from the same
quality of pig-iron, but with spiegel (the German metal) added in
the first case, and ferro-manganese—iron, carbon, and manganese
—in the second :

	With Spiegel.	With Ferro-Manganese.
Limit of elasticity .	22 tons per sq. inch.	16 tons per sq. inch.
Breaking strain . .	32 ,, ,,	28 ,, ,,
Elongation p.c. mea- sured over 8 in. }	8 ,, ,,	25 ,, ,,

"This decrease of the breaking strain, with the increased
elongation," he says, "is a decided advantage where hardness of
the material is not specially required. The metal which withstands
a heavy breakage load with a small final elongation has a special
elasticity in the shape of resistance to change of form, which is
apparent even when it is worked hot. It is necessary, when
steel is to be used for plates, forgings, machinery, and such like
purposes, that it should be very soft. From a constructive point
of view, the exact value of a material is the product obtained by
multiplying the breakage strain by the final stretching, and not
the breakage strains alone." Applying this principle to the two
kinds of steel named above, and also to common iron, he gets
the following result :—

Hard ordinary steel . . . . = 305
Soft steel . . . . = 700
Common iron . . . . = 105

It was this soft steel which the Bessemer process was now
capable of producing. The process, when described by Sir
Henry Bessemer at Cheltenham, in 1856, was so nearly complete

that only two important additions were made afterwards. One was the introduction of the ferro-manganese, just described, for the purpose of imparting to his pure liquid iron the properties of mild steel. The other was an improvement in the mechanical apparatus. He found that when the air had been blown into the iron till all the carbon was expelled, the continuance of "the blow" afterward consumed the iron at a very rapid rate, and a great loss of iron thus took place. It was therefore necessary to cease blowing at a particular moment. At first he saw no practical way by which he could prevent the metal going into the air-holes in the bottom of the vessel below the level of the liquid mass, so as to stop them up immediately on ceasing to force the air through them ; for if he withdrew the pressure of air the whole apparatus would be destroyed for a time. Here, again, his inventive genius found a remedy. He had the converter holding the molten iron mounted on an axis, which enabled him at any moment he liked to turn it round and to bring the holes above the level of the metal ; whenever this was done the process of conversion or combustion ceased of itself, and the apparatus had only to be turned back again in order to resume the operation. This turning on an axis of a furnace weighing eleven tons, and containing five tons of liquid metal at a temperature scarcely approachable, was a system entirely different from anything that had preceded it ; for it he took out what he considered one of his most important patents ; "and," he says, "I am vain enough to believe that so long as my process lasts, the motion of the vessel (containing the fluid) on its axis will be retained as an absolute necessity for any form which the process may take at any future time." The patent for this invention was taken out about four years after his original patent for the converter.

The Bessemer process was now perfect. Nearly four years had elapsed since its conception and first application ; and in addition to the necessary labour and anxiety he had experienced, no less than £20,000 had been expended in making experiments that were necessary to complete its success. It only remained to bring the process into general use, and this was the work that next engaged his attention.

His first step was to get an adequate supply of iron suitable for the converter. In order, therefore, to ascertain whether iron low in phosphorus could be made in England as well as in Sweden, he got analyses of all the different kinds of iron ore raised in the United Kingdom, and finding that the purest and lowest in phosphorus was the hematite of Cumberland, he ordered some pig-iron from Workington and tried it in the converter. Again he was disappointed, for it was as bad as the inferior qualities which he had previously discarded. Analysis showed, to his surprise, that the reason of this failure was the presence of more phosphorus in the iron than there was in the ore. This led him to suspect that there was something wrong in the manufacture of the crude iron. He therefore wrote to the manager and directors of the ironworks at Workington, asking permission to examine their whole system of iron-making, and holding out a prospect of thereby being able to find out something that might benefit both him and them. The directors invited him to dinner, and after discussing the object of their meeting the manager showed him all over the works and explained every detail of their operations. But the keen eye of Sir Henry could detect nothing wrong—nothing that would account for the presence of more phosphorous in the iron than was in the ore. He therefore gave up the search as hopeless, and in company with the manager was crossing the yard from the works on his way back to the dining-room where the directors were still sitting, when he observed a heap of matter in a corner and asked the manager what it was. "Oh, it's nothing worth noticing," was the reply; "it's only mill cinder that we get from Staffordshire and use as a flux." Sir Henry insisted on looking at it, and on examination exclaimed, "Oh, here it is; this is what gives the additional phosphorus to the iron." Satisfied that he was right, he entered the directors' room and with an air of triumph said, " We've caught the villain ; the secret's out now ; it's that mill cinder that does the mischief to your iron." He then advised them to abandon that material as a flux, and asked them to make for him 100 tons of iron without using that flux. The order was duly executed, and when the iron so made was put into the converter Sir Henry was delighted to find that it worked admirably. This iron was marked B, and was the first of that quality which is now universally known as Bessemer pig.

His next task was to convince the public that an invention which for two or three years had been entombed in the oblivion of demonstrated failure was now a complete success. To do this required the exercise of more than ordinary skill and courage. The incredulity with which great discoveries and inventions have almost invariably been received by the public, when viewed through the perspective of subsequent events, forms one of the most remarkable chapters of human history. The execrations of ages have been poured upon those who became the enemies of Galileo because he propounded the diurnal rotation of the earth ; nevertheless succeeding generations have rarely failed to exhibit a like spirit of incredulity, though in a less violent form. "When one considers the splendour of Newton's discoveries," observes one of the greatest admirers of the prince of science, "the beauty, the simplicity, and grandeur of the system they unfolded, and the demonstrative evidence by which that system was supported, one could hardly doubt that to be received it required only to be made known, and that the establishment of the Newtonian philosophy all over Europe would very quickly have followed the publication of it." Yet, incredible fact ! it was not till thirty years after the publication of the *Principia,* which eventually effected an entire revolution in mechanics, that its discoveries could be smuggled into Cambridge University. To the succeeding generation Dr. Samuel Clarke issued a new translation of the French book which was then recognised as the authoritative expounder of the old philosophy; and to it he appended notes which explained the views of Newton, and which, while avoiding all appearance of controversy, refuted the text. This stratagem completely succeeded. The truth supplanted error without alarming prejudice or awakening from its lethargy the dread of innovation. In every age there are people who think themselves interested in maintaining the existing state of things. Sir Henry Bessemer knew this only too well. In the present case his difficulty was aggravated by the recollection of previous failure. However, he could not afford to wait till the obliterating hand of time had disarmed prejudice. To recoup himself for the thousands of pounds and years of labour he had spent in working out his invention, its immediate adoption was necessary; and accordingly he confidently but cautiously proceeded to put it into practical operation.

Having converted the hematite iron, henceforth to be called Bessemer iron, into steel by his process, he wished to demonstrate its properties by actual use. With this view he asked his friend, Mr. Galloway, of Manchester, to distribute the new metal among his workmen when they asked for steel to make tools with, but not to let them know that it was in any way different from what they had been accustomed to use. This was done. The steel was distributed in the usual way to make tools with, and in six weeks Sir Henry Bessemer returned to Manchester to hear what was the result. " What do the workmen say about the new steel ? " inquired the anxious inventor concerning the first product of his infant industry. " They have said nothing at all about it," replied Mr. Galloway. " Nothing at all ! Oh, then, it will be all right; if they have no fault to find with it that is the best report of any." Not content, however, with this silent commendation, Sir Henry went round among a few of the workmen, and in course of conversation asked what they thought of the steel they had got last. The first reply to his question was, "There's no difference between it and other steel; it's no better than we used to get." Such a recommendation was sufficient. The steel formerly used cost 60*l.* a ton; this new steel cost 6*l.* or 8*l.* a ton.

More than ever confident in his success, Sir Henry Bessemer brought the merits of his process before the Institution of Civil Engineers in a paper which he read on the 24th of May, 1859. He then explained that in the three years that had elapsed since he brought the subject before the British Association he had pursued one undeviating course, having determined to remain silent for years under the expressed doubts of those who predicted the failure of his process, rather than again bring forward the invention until it had been practically and commercially worked, and until there had been produced by it both iron and steel of a quality which could not be surpassed by any iron or steel made by the tedious and expensive process previously in general use. Having explained the difficulties that he had in the interval surmounted, he described the improved process in language which is still considered to give the most graphic account of it. He said: "The converting vessel is mounted on an axis, at or near its centre of gravity. It is constructed of boiler plates, and is lined either with fire-brick, road drift, or ganister (a local name in Sheffield for a peculiar kind of powdered stone) which resists the heat better than any other material yet tried, and has also the advantage of cheapness. The vessel having been heated is brought into the requisite position to receive its charge of melted metal, without either of the tuyeres (or air holes) being below the surface. No action can therefore take place until the vessel is turned up (so that the blast can enter through the tuyeres). The process is thus in an instant brought into full activity, and small, though powerful, jets of air spring upward through the fluid mass. The air expanding in volume divides itself into globules, or bursts violently upwards, carrying with it some hundredweight of fluid metal which again falls into the boiling mass below. Every part of the apparatus trembles under the violent agitation thus produced ; a roaring flame rushes from the mouth of the vessel, and as the process advances it changes its violet colour to orange, and finally to a voluminous pure white flame. The sparks, which at first were large like those of ordinary foundry iron, change into small hissing points, and these gradually give way to soft floating specks of bluish light, as the state of malleable iron is approached. There is no eruption of cinder as in the early experiments, although it is formed during the process; the improved shape of the converter causes it to be retained, and it not only acts beneficially on the metal, but it helps to confine the heat, which during the process has rapidly risen from the comparatively low temperature of melted pig-iron to one vastly greater than the highest known welding heats, by which malleable iron only becomes sufficiently soft to be shaped by the blows of the hammer; but here it becomes perfectly fluid, and even rises so much above the melting-point as to admit of its being poured from the converter into a founder's ladle, and from thence to be transferred to several successive moulds.''

He next exhibited specimens of the metal produced and explained the severe tests which they had stood. Its extreme toughness and extensibility were proved by the bending of cold bars of iron, 3 inches square under the hammer to a close fold, without the smallest perceptible rupture of the metal at any part, though the bar was extended on the outside of the bend from 12 to $16\frac{3}{4}$ inches, and compressed on the inside from 12 to $7\frac{1}{4}$ inches, making a difference in length of $9\frac{1}{2}$ inches between what before bending were the two parallel sides of a bar 3 inches square. He also explained that this metal could be made for 6*l.* a ton.

Notwithstanding numerous proofs of this sort, some of the speakers who addressed the meeting after hearing the paper read, expressed grave doubts as to the regular or practical working of the process.   On the whole, however, the paper produced such a favourable impression as to the ingenuity of the process that the Institution resolved to present to its author the Telford gold medal.

The process had yet to be made a commercial success. "After," he says, "I had succeeded in making steel of so good a quality that it was not recognised as being anything different from the ordinary high Sheffield article, I then brought the subject again before the public, and was surprised to find that no one believed in it; no one seemed to have the smallest confidence in it; every one said: 'Oh, this is the thing which made such a blaze two or three years ago, and which was a failure.'   Had I not been furnished with capital by the sale of licenses my experiments could never have been carried on.   I had acquired five powerful friends—men who would have an advantage of 10,000*l.* a year each over their fellow ironmasters, because they would have no royalty to pay—yet not one of them made the smallest attempt to get over the first difficulties of the process; they simply treated it (to use an expression common at that time) as a meteor that had passed through the metallurgical world, but that had gone out with all its sparks.   I had immense difficulty in persuading any one to touch it; indeed, neither the steel makers nor the iron makers would take it up after the lapse of two years.   I then saw that the thing was hopeless; so I and my partner, Mr. Robert Longsden, joined with the Messrs. Galloway, of Manchester, for the purpose of erecting works for the production of the new steel.   We bought land at Sheffield, and erected works there, with the determination of convincing people that the process was really good, or else, failing in the attempt, to give the thing up.   Of the five parties to whom I had originally given licenses, two of them, who paid me 10,000*l.* each for one year's royalty, spent about 100*l.* or 150*l.* each in trying the process; the other three spent nothing. Now, when I was successful in making steel (the first invention was for iron only), I found that my having sold them this privilege would, if it applied to steel, have given them an advantage of 2*l.* a ton, or altogether 40,000*l.* over other manufacturers. I saw that others could not fairly compete under these conditions, so I then applied myself to repurchase those licenses. I gave one firm 10,000*l.*, and another 20,000*l.*, for the privilege which they had purchased but left unused for five or six years, and for which they had only given me 10,000*l.* originally. I swept the market clear of all those licenses. In due time our works were erected, and we commenced to sell. We sent out one traveller with samples of engineers' tool steel, quoting 42*l.* per ton, a price which we maintained for the whole two years during which this branch of the trade was carried on. So little confidence was there in it then, that at first our orders were for 28 lb. or 56 lb. of steel at a time—most paltry orders. These were, however, all duly executed; they soon became larger, and afterwards very much larger. Then the manufacturers in Sheffield began to say; 'Why, these people are underselling us by 20*l.* a ton.' Sir John Brown, my next door neighbour, was the first man to look into it. He was about to erect most expensive works for producing steel by puddling, intending to remelt the puddled steel in a crucible on Krupp's plan. As soon as he saw the inside of our works he abandoned that idea, and proposed to take a license, which was granted, but not at the original 10*s.* royalty, for I had raised my royalty to 2*l.* a ton on all articles, except rails, and 1*l.* a ton on rails, taking in that case only a small portion of the saving on those articles."

The cheapness of the new steel was not the only quality that recommended it. Shortly after the erection of the Bessemer

Steel Works at Sheffield, the inventor was visited by Mr. Parks of Birmingham, who had invented a system of making copper tubes by pressing a round disc through a hole with a plunger, thus producing a kind of cup, which was subsequently extended by drawing in dies to a tube with a closed end. Mr. Parks said he believed that the steel produced by the Bessemer process would act in the same way as the best copper, which he was obliged to use for that purpose. Sir Henry Bessemer said: "I don't believe anything of the kind; it seems to me utterly impossible that so rigid a material as steel could safely undergo such a process." But Mr. Parks was so sanguine that it could be done that he induced Sir Henry to go with him by train to Birmingham that night, taking a disc of steel to try if this could really be done. The disc in question measured 23 inches in diameter and $\frac{3}{4}$ of an inch thick, and was cut from the end of a locomotive tube plate of mild steel, which the Bessemer firm was then manufacturing to the order of the Lancashire and Yorkshire Railway Company. Mr. Parks placed this steel disc on the top of a cast-iron die, with a hole in it 11 inches in diameter. The mouth of the die was in shape like that of a French horn. When he had forced the cold steel disc half way through the die, Mr. Parks said: "We will do with this what we always have to do with our copper." It was then taken out of the die, put into the annealing furnace, and left to get cold again. On being put into the press a second time, the steel was entirely pushed through the die. Thus a steel plate $\frac{3}{4}$ of an inch thick was formed into a deep cup without the smallest injury to the metal. Rigid as it was, says Sir Henry, this plate of 23 inches in diameter had by some magic movement of its particles been reduced to 11 inches in diameter, and its flat surface changed to a cup of 10 inches in depth. He himself was so astounded at this proof of the quality of his steel that he has preserved that cup as a memorial of the event.

Thus the new process began to make progress. Its superiority, in respect of rapidity and cheapness over the old process, was spreading consternation in the steel trade. One manufacturer after another applied for a license to use it. Others endeavoured to secure its advantages by other means. " One day," says Sir Henry, " I found in London a gentleman occupied in his office with a packet of papers a foot high before him, getting out all the cases he could against me for repealing by *scire fascias* the whole of my patents. He was employed by a company of ironmasters to do so, and he told me candidly enough afterwards : ' When I had gone through the whole of your patents, and about seventy patents which they said more or less anticipated yours, I found that they had not a leg to stand upon, and I advised them to come to you for a license.' As there had been a great deal of scurrilous writing against me by one of the parties connected with the firm, I said, when they applied for a license, ' I know you only come now for a license because you cannot upset the whole of my patents ; however, I shall not refuse on that ground, but I refuse it until I have a letter of apology from one of your people, such a letter of apology as will show that the statements made against me were without foundation. The moment I get such an apology as one gentleman should give to another I will give you a license to manufacture ; but I will never deal with a man who has written against me in that way.' I received an entire re-tractation, a most perfect and gentlemanly apology, and I then granted a license to that company, but they have never used it."

The process being now in operation at Sheffield, Sir Henry Bessemer gave an account of its results to the Institution of Mechanical Engineers when they met in that town in August, 1861. In conclusion he used these memorable words : " For the practical engineer enough has already been said to show how important is the application of cast steel to constructive purposes, and how this valuable material may be both cast and forged with such facility, and at a cost so moderate, as to produce by its superior durability and extreme lightness an economy in its use as compared with iron   The construction of cast steel girders and bridges, and of marine engine shafts, cranks, screw propellers, anchors, and railway wheels, are all deserving of careful attention. The manufacturer of cast steel has only to produce at a moderate cost the various qualities of steel required for constructive purposes to ensure its rapid introduction ; for *as certainly as the age of iron superseded that of bronze, so will the age of steel succeed that of iron.*"

CHAPTER IV.

IN the history of inventions there are many instances in which the discoverers of principles, that formed the germ of revolutions in science and art, allowed their great ideas to lie dormant till some more enterprising spirit appropriated them, and boldly commanded public attention to their merits in order that they might reap the reward of their application. This was not the case with the Bessemer process. As yet, however, the age of steel was only a potentiality, not a reality. The inventor was now reaping handsome profits from his invention; but it had not yet become a benefit to the community at large. Its success was practically established, but its usefulness was not generally appreciated. The application of the new steel to industrial purposes had yet to be accomplished, and Sir Henry Bessemer found that this was not the least formidable part of the work set before him. The manufacturers of iron and steel henceforth showed increasing eagerness to appropriate such a profitable invention; but the consumers of iron and steel had yet to be convinced that a new metal so easily produced was the best adapted for industrial purposes.

In a leading technical journal [1] of the period (1862), we find it stated that "To produce a metal possessing such superior

[1] *The Engineer.*

qualities from English pig-iron in twenty minutes is an achievement which, in 1851, would have been thought miraculous. Now it is an everyday performance at some of the greatest steel works in this country and on the Continent. Those who believe that every great invention is, like the steam-engine, the result of accumulated improvements, all in the same direction, may well regard the Bessemer process with interest. Mechanically, although not commercially, it has already effected a complete revolution in the manufacture of iron and steel, so much so, that those who have adopted it are receiving 40*l.* and 50*l.* a ton for steel which costs them less than 10*l.*, and which would successfully compete, even at the same price, with steel quoted at 60*l.* or 70*l.* Yet the process of converting crude melted iron into steel by simply blowing air through it for a quarter of an hour or twenty minutes, and without any fuel other than that first required to melt iron, was, as far as the past six years have shown, the sole conception of one man. It is wonderful that, with such a great accomplished fact, his right to the whole process has never been disproved and never disputed."

At the Great Exhibition of 1862 he was an exhibitor of numerous specimens of his steel, manufactured into a variety of articles, ranging from heavy steel ordnance to steel wire the two hundred and fiftieth part of an inch in diameter. Specimens made by the same process were also shown that had been sent by companies in India, Sweden, France, Germany, and Austria, where Bessemer steel was then being produced.

The history of the process was as eventful abroad as at home. A fortnight after the first announcement of the process in England it was reported that its inventor had realised 80,000*l.* by the sale of his patent rights abroad, so great was the first sensation caused by it. Speculators even negotiated the purchase of the patent rights for Spain—the least industrial country in Europe. On September 26, 1856, a trial of the process was made at

Baxter House for the express purpose of satisfying the gentlemen who wished to purchase the Spanish patent. The price was £5,000, and the trial was successful. But the discredit into which it fell immediately afterwards soon spread abroad ; and the disappointments experienced by Sir Henry in consequence turned out to be more irremediable abroad than at home.

Amid these disappointments and failures Sweden was the one bright spot where it failed not. That country was the first to adopt the Bessemer process, and it did so soon after the first announcement of its success. The exceptional purity of the iron made in that country rendered it specially suitable for the converter. Accordingly, a leading manufacturer there applied to the inventor for a license to use it ; and the original form of converter, without the later improvements, was used not only at first but for years afterwards. So great was the interest excited by the introduction of the process there, that the Crown Prince, who was President of the Iron Board of Sweden, inspected the first operation of making steel, and he was so satisfied with it that he made the inventor an honorary member of the Iron Board. The first Bessemer steel ingot ever rolled in England came from Sweden. It was rolled into a circular saw plate of five feet diameter, and was preserved by Sir Henry Bessemer as the first steel article produced by precisely the same apparatus, and the same method of treatment, that he described at Cheltenham in 1856.

In Germany the greatest steel maker was Herr Krupp of Essen. Before Sir Henry Bessemer had taken the preliminary steps to obtain a patent there, Krupp entered into negotiations with him and agreed to pay £5,000 for the use of his invention. This license was for the whole of Prussia, and practically gave him the rights of the inventor in that country, so that Krupp could either have made the process a monopoly at his own works, or could have leased it to others. The inventor, accordingly,

G

sent all his papers to Krupp, who, in due course, applied to the Prussian Government for a patent. The Prussian Government told Krupp that the invention was not a new one. He pressed them to show who had done it before, and they named Mr. Nasmyth as having made the invention previously. Mr. Nasmyth denied having done so, and when this was represented to the Prussian Commissioners of Patents, they said then that some one else had done it, and they would find out who it was in a few days. A few days passed, and the Prussian commissioners being still unable to find their "somebody else," promised to do it in another few days. Thus, they said, they continued their search, always maintaining that it was an old invention which would be very soon found out. Six weeks passed without it being found out, and they then began to promise the importunate Krupp to find it day by day. At last they said: "If we do not find it to-morrow, we will give it you." So this inquiry went on from one to-morrow to another, until, as Sir Henry puts it, there was a week of to-morrows. On the last occasion of his calling they presented Krupp with an English blue book, containing the publication of the English patent, and said : "Now, seeing it is a publication in Prussia, we cannot grant you a patent by the law of Prussia." It is scarcely necessary to add that the process was worked in Germany without payment of any royalty to the inventor.

This failure to secure the profit of his invention in Germany led to a similar result in Belgium. Sir Henry obtained a patent in Belgium, but when Krupp of Essen sent his Bessemer steel into that country, the Belgian manufacturers, who had agreed to pay the inventor for licenses, represented to him that they could not continue the payment, because, while doing so, they were unable to compete with the same description of manufactures that came from Essen. The directors of the Seraing Company determined to ascertain whether the steel in question was made

by the Bessemer process at Essen, and for this purpose directed
two of their workmen who were well informed in the details of
the process to represent themselves at the Essen works as skilled
Bessemer steel makers in search of employment, and as prepared
to impart a knowledge of some special features of the process
peculiar to Belgium. Their services were accepted, and they had
thus an opportunity of witnessing the working of the Bessemer
process in the Essen Works. On their return with this information the directors of Seraing refused to pay the royalty on the
process any longer, and the Belgian patent of the inventor was
henceforth of no value.

The story of the working of the Bessemer process in the
largest works in France affords another striking illustration of the
genius of its inventor, and of the ingenuity of foreign manufacturers in appropriating it. At the meeting of the Iron and Steel
Institute in London, in 1877, Mons. Gautier, of Paris, read a
paper on "Solid Steel Castings." He explained that "when
steel is cast in an iron ingot mould, or a mould of any kind,
usually the metal after cooling is not entirely sound. Cavities of
a more or less rounded shape are seen inside, apparently caused
by a gas escaping from the mass. Sir Henry Bessemer, the first
among metallurgists, has demonstrated that these blow-holes
were filled with oxide of carbon (carbon combined with oxygen),
and this view has since been entirely confirmed. When these
blow-holes are altogether inside, and do not burst through the
crust, they remain silvery white, and to get rid of them it is
sufficient to weld the metallic parts by re-heating, and the use of
the hammer or rolls. When the blow-holes communicate with
the outside, and the sides of the ingots are pierced with small
holes, well known to the steel manufacturers, the colour is no
longer a silvery white; they assume more or less the colours of the
rainbow, and even become black. . . . It is an easy matter to
remove blow-holes when the steel has to undergo mechanical

elaboration, but it is not so with castings, and it is very important to prevent the formation of these blow-holes when sound pieces are wanted, and pieces the resistance of which can be relied on. . . . . The German manufacturers have produced ingots without blow-holes, beginning with two tons weight and finally reaching 45 tons. The Krupp ingots and the cast wheels and bells of Bochum certainly astonished the metallurgical world for some time. The process of manufacture was kept a most profound secret, and has not yet been published. More than six years ago the Terre Noire Steel Works found out, by reasoning rather than by practice, the process of the German works, and the improvements they have made have radically transformed the result. It seems well proved now that the German products without blow-holes are obtained by the addition of a very silicious iron just before casting." Mons. Gautier related some facts to show that this cast metal was superior to forged steel both in strength and regularity, and concluded by saying that the industrial applications of this metal would naturally present themselves to every mind, and would push themselves forward, bearing with them these two great advantages, solidity and economy.

In the discussion that followed the reading of this paper, Sir Henry Bessemer said he observed in the process just described an old friend come back to head-quarters ; and he proceeded to explain that in the Exhibition of 1862 he exhibited a number of castings, chiefly railway crossings, among them being an ingot of about 17 cwt., one half of which was turned down in the lathe, and the end of it " faced," to show that not a single blow-hole existed in it. That ingot and those castings were produced by the use of silicon (from *silex*—flint) and manganese. He knew well that one of the defects of his process was that occluded gases constantly produced air bubbles in the ingots and castings, and it was the discovery of a mode of getting rid of them that induced him to make those samples and show them at the

International Exhibition of 1862. He then gave an account of the circumstances that led to that discovery. When he first went down to Sheffield, about 1856, he was a stranger to the ordinary mode of producing cast steel. He heard steel makers say that some of their ingots worked successfully and were perfectly sound, while others were full of air bubbles and worked very badly. The terms used in the trade to distinguish these two characters of steel were "well melted" and "not well melted." He observed that these manufacturers usually put their best Swedish iron, or blistered steel, into the crucible with a little manganese and carbon in the form of charcoal. If they allowed it to remain only a sufficient time to fuse the metal, the result was a bubbly ingot that went to pieces under the hammer ; but if they allowed it to remain an hour or more after being melted, and increased the the temperature of the crucible, it became what was called " well melted," and worked perfectly well. He determined to know what was the chemical difference between these metals, and an examination made, with the assistance of some able chemists, of six or seven samples of each kind of steel produced from the same iron, showed him that the "well melted " steel contained a small quantity of silicon, which was entirely absent in the iron from which these ingots were made. He thought, therefore, that the silicon must produce the superior quality of steel, but was at a loss to know how the manufacturers got this appreciable quantity of silicon into the first class Swedish bar iron of which the steel was made. Further observation gave him a key to this question. He found that in making the clay crucibles a plug was formed at the bottom of the core used in the moulds in order to make the core go into its right place. The plug necessarily made a hole of about an inch in diameter in the bottom of the crucible. If they were to attempt to patch up or stop this opening with moist clay, the shrinkage of the clay plug in drying would make the crucible very leaky, and to prevent this they placed the crucible on a small fire-lump and threw a handful of sand into it in order to stop up the hole and to prevent oxidation of the metal. They also put into the crucible a lump or two of charcoal. It followed that when they allowed the metal to be merely melted in the crucible, and then poured it out at once without allowing it to acquire a very high temperature, probably none of the silica (sand) was converted into silicon, and in that case they obtained a bubbly ingot; but when they allowed it an hour or two more and increased the temperature, a quantity of the silica and manganese was reduced and formed silicon, producing "well melted" steel which was free from air bubbles, and which worked well. Though no steel-maker in Sheffield appeared to know why it was that he made good steel one day and bad another, Sir Henry satisfied himself that he had found the true explanation. When, therefore, his own process was put into practical operation, he searched among the various pig-irons of this country to find those that would give him manganese and silicon together. For this purpose he went to Tow Law, in Durham, and tried the pig-iron made from spathose ore in use there. On analysis he found that it contained $4\frac{3}{4}$ and sometimes 5 per cent. of silicon, and about 3 or 4 per cent. of manganese. He began at once to employ that material. Having, by blowing in the converter, reduced this grey hematite pig-iron to the state of soft malleable iron, he used this alloy of silicon in addition to spiegeleisen, and by so doing he at once obtained "well melted" steel, which ceased to boil and bubble in the mould, and from which perfectly sound castings were produced at his works. "It is a remarkable fact," said Sir Henry, "that that which a French metallurgist brought before the Iron and Steel Institute in 1877 as a new discovery, had been practised for fifteen years in the Bessemer Steel Works at Sheffield."

But this was not the only remarkable feature of the case. Among the persons who visited the Bessemer Steel Works shortly after this discovery was made, for the purpose of learning all the details of the Bessemer process, were the Messrs. Schneider of Creusot, the owners of the largest works in France. There was a good deal of doubt among French metallurgists as to whether the process was capable of converting French iron into steel. In 1863, however, Mr. Schneider's son came to Sheffield, and arranged with Sir Henry Bessemer for a license to make steel at Creusot by the new process. Terms were settled—he signed the deed of license—and Sir Henry gave him eleven sheets of detailed drawings for the erection of works, the construction of converters, and so on. Their manager also passed a month at the Sheffield works for the purpose of being thoroughly initiated into all the details of the process. To make sure that the pig-iron made at Creusot was suitable for the Bessemer converter, four tons of it were sent over to Sheffield, and were converted into steel ingots in the presence of Mr. Schneider, his son, and his manager. The steel thus made was worked and hammered into bars, which they took back with them to France. They were, they said, perfectly satisfied that they could make steel from their own iron by the Bessemer process. They were afterwards occupied for about two years in making a huge workshop full of machinery on the plans and drawings that Sir Henry had supplied. Nothing was charged by him for those plans or the right to manufacture his patent machinery, as the remuneration was to be paid entirely on the tonnage of steel produced. From these plans the largest steel works in France were set up at Creusot, and were finished just a few weeks before the expiration of the inventor's French patent. They were not, however, set to work during these few weeks, but immediately the patent had expired the works commenced to make steel in large quantities. Not only did they refuse to pay the inventor a penny of royalty for the use of the new process or for the special instructions and drawings received from him, but the small royalty for the improved apparatus for turning the converter up and down, the patent for which had not yet expired, was repudiated. Following the example of the Messrs. Schneider, the Terre Noire Company also refused to pay any royalty. "And now," said Sir Henry, in 1877, "these French gentlemen are kind enough to tell us in England how to make solid steel castings."

While the inventor thus saw one foreign steel manufacturer after another adopting his process and eluding payment of his royalty, he had the satisfaction of seeing its steady progress in England, as well as on the Continent and in the United States.

In 1865 he stated that there were then "seventeen extensive Bessemer steel works in Great Britain. At the works of the Barrow Steel Company 1,200 tons per week of finished steel can easily be turned out, and when their new converting house, containing twelve more five-ton converters, is completed, these magnificent works will be capable of producing weekly from 2,000 to 2,400 tons of cast steel. There are at present erected and in course of erection in England no less than sixty converting vessels, each capable of producing from three to ten tons at a single charge. When in regular operation these vessels are capable of producing fully 6,000 tons of steel weekly, or equal to fifteen times the entire production of cast steel in Great Britain before the introduction of the Bessemer process. The average selling price of this steel is at least 20*l.* per ton below the average price at which cast steel was sold at the period mentioned. With the present means of production, therefore, a saving of no less than 6,240,000*l.* per annum may be effected in Great Britain alone, even in this infant state of the Bessemer steel manufacture."

Sir W. Siemens has called Sir Henry Bessemer "the pioneer of steel for structural purposes." But his title to that distinction was not easily earned. For years he had to wage continual warfare against time-honoured prejudices, trade interests, and stubborn ignorance. The Government of the country which has the honour of being the birthplace of the process, and which has been most benefited by it, was the first and the last to resist its adoption.

It will be remembered that the original intention of Sir Henry Bessemer was to produce a better metal than ordinary iron for ordnance. Having succeeded in this, he first brought his new metal under the notice of the authorities at Woolwich Arsenal. He not only informed Colonel Wilmot, who was then super-intendent at Woolwich, of the successive steps he had made in the progress of his process, but invited him to visit his new works at Sheffield in order to see it in practical operation. Colonel Wilmot readily availed himself of the opportunity. He went to the Bessemer Steel Works at Sheffield, and there studied the process and its products, and thus satisfied his mind that steel was the material for the future guns of the country. Shortly after that Sir Henry read his paper before the Institution of Civil Engineers—that paper which was rewarded with the Telford gold medal—in the course of which he said, with reference to the production of ordnance from his new metal, that "in order to show the extreme toughness of such iron, and to what a strain it may be subjected without bursting, several cast and hammered cylinders were placed cold under the steam-hammer, and were crushed down without the least appearance of tearing the metal. Now, these cylinders were drawn down from a round cast ingot, only two inches larger in diameter than the finished cylinder, and in the precise manner in which a gun would be treated. They may, therefore, be considered as short sections of an ordinary 30-pounder field-gun. Iron so made requires very little forging, indeed the mere closing of the pores of the metal seems all that is necessary. The tensile strength of the samples, as tested at the Royal Arsenal, was 64,566 lbs., or nearly thirty tons, per square inch, while the tensile strength of pieces cut from the

Mersey gun gave a mean of 50,624 lbs. longitudinally and 43,339 lbs. across the grain, thus showing a mean of 17,550 lbs. per square inch in favour of the Bessemer iron. If it be desired to produce ordnance by merely founding the metal, then the ordinary casting process may be employed with the simple difference that the iron, instead of running direct from the melting furnace into the mould, must first be run into the converting vessel, where in from ten to twenty minutes it will become steel, or malleable iron if desired, and the casting may then take place in the ordinary way. The small piece of ordnance exhibited will serve to illustrate this important manufacture, and is interesting in consequence of its being the first gun that was ever made of malleable iron without a weld or joint. The importance of this fact will be much enhanced when it is known that conical masses of this pure, tough metal, of from five to ten tons in weight, can be produced at Woolwich at a cost not exceeding 6*l.* 12*s.* per ton, inclusive of the cost of pig-iron, carriage, re-melting, waste in the process, labour, and engine-power. These facts have been laid before the Government, and their advantages are fully appreciated by Colonel Eardley Wilmot, the superintendent of the Royal Gun Factories, who has evinced a great interest in the progress of the invention from its earliest date, and to whose kindness the author is indebted for the many valuable trials of the tensile strength of the various samples of metal that have been submitted for investigation."

In the discussion that followed Colonel Wilmot said he had from the commencement of these inquiries taken a great interest in them, and had mechanically tested the products originally produced. As regarded the difficulties of the process as well as the results of it, he thought that the best thing for a member of a practical society to do was to follow his example, and go and see it for himself. Nothing could be more simple or more perfectly under control, and having by a few trials ascertained the particular kind of treatment required with the samples of iron to be dealt with, it was operated upon with certainty. After giving some evidences of the superiority of Bessemer steel, he mentioned that he had been using it for turning the outside of iron guns, cutting off large shavings several inches in length, and he found none superior to it, although much that was more costly. It was only necessary to witness the operation of the Bessemer process to be satisfied that the expense of converting the pig-iron into any of its products involved scarcely any cost beyond the labour, and that but for a very short period of time.

Prof. Abel, the director of the chemical department at Woolwich, who had always taken a lively interest in the new metal, likewise said, in 1860, that there appeared no reason why the Bessemer process, "which has recently been applied with great success to the conversion of iron of good chemical quality into excellent cast steel, upon a very considerable scale, should not be resorted to for the production, at a moderate cost, of masses of cast steel, or a material of a similar character, of sufficient size for conversion into cannon of all sizes but those of the heaviest calibre, which it will always be found most advantageous to construct of several pieces."

Colonel Eardley Wilmot being naturally anxious to put the new process in operation at Woolwich, invited Sir Henry Bessemer to go there with him to inspect the foundry department with the view of adapting it for the process. Sir Henry went, and as the result of their deliberations, it was decided to remove two large furnaces to make room for his apparatus. All the preliminary arrangements being made, Sir Henry was asked to prepare plans and an estimate, and these having been duly lodged at Woolwich, he was in daily expectation of their passing through the necessary red-tape ordeal so that he could commence to put in his apparatus. Having no answer for a week or ten days, he was surprised, because he had been told that the matter would be put through instantly. He therefore went down to Woolwich, but only to find that in the interim Colonel Wilmot had been superseded. Proceeding to inquire what was to be done next, he saw Mr. Sidney Herbert on the subject, and explained the matter to him at very great length, endeavouring "to coach" him upon the merits and utility of his steel. Mr. Sidney Herbert listened to him attentively, and said in reply that it was a technical question which he did not understand, but he would investigate the matter. Sir Henry said: "Then you have only to see Colonel Wilmot, who has for the last three years given great attention to the subject; he has seen the process over and over again, and has tested it in every way that he could test it. Besides Colonel Wilmot, there is the Laboratory Department, where the chemical analyses can be obtained, and the testing machines will show you what it is; there is also the rolling-mill to show how the material behaves under the rolls and under the hammer; and then, in addition to that, there are some dozen or two of officials at Woolwich who have witnessed these experiments, and can give you every possible detail." He arranged to see Mr. Sidney Herbert a week later, and went with all the confidence of one who was to have the contract put into his hands. But his hopes were soon dispelled. Mr. Sidney Herbert told him that he had consulted Sir William Armstrong, and understood from him that the new material was wholly unsuited to the manufacture of ordnance. Sir Henry was amazed. Of course he remonstrated. He said: "Why go to my rival? Why go to the gentleman who has a scheme of his own to carry out instead of mine?" Mr. Sidney Herbert replied that he had taken the best advice and had come to the conclusion that he was not to have the contract, and was not to put up his process at Woolwich—they would in future make the guns of iron.

"I left," said Sir Henry, a quarter of a century afterwards, "I left, like many other men who had been at Woolwich, in great and deep disgust. I had, however, too good a business at that time in the introduction of my process to make it worth while to pursue it through all the tortuous paths that one must go through to oppose the Government, so I let them alone. But see what the upshot has been. In the coils to make the bore of the iron gun there were numberless little cracks, so what did they do? They put a steel tube in, and when they got it into the heart and soul of the machine, they tied this steel with a piece of iron. Why not with a piece of steel? Why not a cylinder made of steel in preference to that iron coil? For the making of steel cylinders was then an accomplished fact, but the making of those iron coils was not an accomplished fact. The iron coil system has been thoroughly shown up, but at an enormous expense to this country. This incident was the turning point which made us have iron guns, while every other country in the world has got steel guns."

In 1882, the year after Sir Henry made these remarks, the British Government determined that the manufacture of wrought iron guns should be discontinued, and that all guns in future should be made wholly of steel.

After the rejection of his offer by the British Government, Sir Henry turned his attention more energetically than ever to the use of steel for industrial purposes. While the production of high class steel from foreign iron was going on at the Sheffield works, no efforts, he says, were spared to introduce steel made from Cumberland hematite Bessemer pig for railway and general engineering purposes. Its homogeneous character and great toughness marked it out as the material for the rails of the future —an article hitherto largely made from the lowest of all grades of iron. This fact was eagerly seized upon by those who wished to disparage the new steel. "Oh, yes," they said, "it may do for rails—anything will make a rail." Nor was science or art less sceptical. In 1861 he proposed to an eminent practical engineer,

Mr. John Ramsbottom, of the London and North Western Railway, to try the use of steel instead of iron rails. At the very mention of steel rails, Mr Ramsbottom, looking at him with astonishment, exclaimed : " Mr. Bessemer, do you wish to see me tried for manslaughter ! " Undismayed, the sanguine inventor showed the amazed engineer some samples of the rails he had made ; and after hearing their mode of production and superior properties explained, Mr. Ramsbottom said : " Well, let me have ten tons of this material that I may torture it to my heart's content." The order was excuted, Mr. Ramsbottom had the steel severely tested, and rolled some of it into rails. These rails were exhibited at the Great Exhibition of 1862 ; and they were afterwards put down at Crewe on the day the Prince of Wales was married. Speaking of these same rails in 1881, Mr. Webb, chief of the locomotive and rail works of the London and North Western Railway Company, stated that they had remained there ever since with the exception of ten months when he borrowed them to send to South Kensington ; and the bulk of the "up" traffic through Crewe Station passed over them. In 1881 the second head was fairly worn down, but they were still in use. Formerly they had to turn iron rails there about every nine months.

At the Birmingham meeting of the British Association in 1865, Sir Henry Bessemer explained that at Chalk Farm steel rails were laid down on one side of the line and iron rails on the other, so that every engine and carriage there had to pass over both steel and iron rails at the same time. When the first face was worn off an iron rail it was turned the other way upwards, and when the second face was worn out it was replaced by a new iron rail. When Sir Henry exhibited one of these steel rails at Birmingham only one face of it was nearly worn out, while on the opposite side of the line eleven iron rails had in the same time been worn out on both faces. It thus appeared that one steel rail was capable of doing the work of twenty-three iron ones.

Perhaps in no department of industry has the application of the Bessemer process caused a greater or more permanent improvement. The superiority of steel to iron rails is not now questioned. It is believed that on the average a steel rail lasts nine times as long as an iron one, and the difference in cost between the two is now trifling. In 1880 about 16,000 miles of railway in Great Britain had been laid with steel; and it is estimated that when the whole railway system of this country, 25,000 miles, is relaid with steel, there will be a saving of nearly 3,000,000*l.* a year in the cost of renewals of rails. Were this economy extended to the whole railway system of the world, the annual saving would be over 20,000,000*l.*

In 1863 Mr. Ramsbottom had the first locomotive boiler made of Bessemer steel plates. The makers found it difficult to get all the materials they required from one firm, so they got part from one maker and part from another. This locomotive boiler lasted till 1879—a much longer time than the average life of an iron boiler.

Mr. Webb, of Crewe, claims for the London and North Western Company the credit of having been the first great railway that recognised the importance of the Bessemer process. Steel, he said, has been substituted in nearly every portion of the locomotive which formerly was made of iron. In 1882 the Company had 1,679 engines with steel boilers, and they had every reason to be satisfied with the result. The Company was also the first to use Bessemer steel plates for its passenger vessels. They had the misfortune in 1881 to get one of their steel vessels on a sunken rock at the entrance to Carlingford Loch; and he felt certain that if it had been built of iron it would have become a total wreck. As it was, ninety feet of her keel passed over the sunken rock, which bulged it in some places to the extent of five or six inches, but there was not a single crack in the plates, and no water got into the vessel. Notwithstanding improvements in material, the quantity of rails annually required for repairs and renewals on the London and North Western Railway was 20,000 tons in 1882, when there were 1,770 miles open. For every mile run, the actual loss of rails was about one third of a pound of steel, so that on the London and North Western 15 cwt. of steel disappeared from the rails every hour of the day.

The use of Bessemer steel for the construction of boilers was first adopted in 1860 by Mr. Daniel Adamson of Hyde—one of the largest and most successful boiler makers in England. He visited the Bessemer Steel Works at Sheffield, and was so convinced as to the superiority of the new metal in comparison with the best qualities of iron that he made some boilers of it for Messrs. Platt, of Oldham. These boilers were subjected to severe tests, and were worked at very high pressure. They answered their purpose admirably. They not only showed superior strength, but they effected a saving of fuel, because the steel plates, being made thinner than iron ones, transmitted the heat more rapidly. The suitable nature of steel for boilers, however, continued to be a disputed question, notwithstanding that it had such early and able advocates as Sir Henry and Mr. Adamson. As showing the conflict between opinion and experience, it is worth mentioning that in 1881 a paper was read before the Institution of Naval Architects on the peculiarities in the behaviour of steel plates, and it presented a rather alarming view of the subject. Among those who heard it was Sir Henry Bessemer, who immediately took the trouble to work out some of the results that were to be inferred from the statements and figures given in that paper. According to the rate of corrosion therein attributed to steel plates, Sir Henry found that a boiler made of $\frac{3}{8}$-inch plate would lose so many grains per inch per week that at the end of eight years and a half there would remain only the coat of paint on the outside, not one single grain of metal being left after that period. So alarming a result led him to inquire whether there was any evidence tending to prove such a statement.   He remembered that Mr. Richardson, the manager for Platt Brothers, of Oldham, had been so well satisfied with the first trials made with steel for boilers that he ordered fifty tons of Bessemer steel in order to make six boilers each 30 feet in length by 6 feet 6 inches in diameter, with flues of 3 feet 10 inches running through them, and with a thickness of metal not exceeding $\frac{5}{16}$ of an inch.   Fully twenty-one years having elapsed since that experiment was made, he thought it might afford a practical proof of the duration of steel for boilers. He accordingly telegraphed to Mr. Richardson asking what state these six boilers were in, if they still existed.   The reply was that all the six boilers were at work, and were still in a most satisfactory condition, showing no signs of corrosion.

Next to railways, shipbuilding is the industry which is the largest consumer of iron.   The history of the introduction of iron instead of wood for shipbuilding has often been written, but a full account of the introduction of steel instead of iron has yet to be written.   There are not a few claimants for the honour of having induced the latter change ; but in this matter also Sir Henry Bessemer's claims are incontestable, though often ignored.   Prior to the advent of Bessemer steel, several small vessels intended for river navigation were made of steel, which being a light material enabled them to steer about in shallow water; but the great cost of the steel then used precluded its application to ocean-going vessels.   In 1862-3 Sir Henry endeavoured to draw the attention of shipbuilders to the superiority of his metal to the common iron then used in ships.   He succeeded in 1863 in inducing one shipbuilder to order enough of steel to make a stern-wheel barge, which was followed by another of the same description.   So satisfied was the shipbuilder with the new material that he persuaded the Humber Steam Packet Company to have a paddle steamer of 377 tons built of Bessemer steel.   This vessel was called the *Cuxhaven*, and was built in 1864.   The next vessel

H

built of the same material was the *Clytemnestra*, a clipper ship of
1251 tons, which might fairly be described as the first ocean-going
ship worthy of the name that was built of steel. Her first voyage
was a memorable one, but was too soon forgotten. She
narrowly escaped destruction in the fearful cyclone that swept
over Calcutta in October, 1864; and the following account of the
perils she went through was taken from her log on her return
to the Mersey:—

" The morning of October 5th, 1864, commenced with strong
winds and thick drizzling rain. At 8 A.M., had gale and
tremendous squalls, with thick constant rain. From 8 A.M. until
noon gale rapidly increasing, and barometer falling fast, with very
threatening appearance. At 2 A.M., tremendous gale and most
terrific squalls, with thick rain and dismal appearance. The ships
attached to the same moorings below us began to break adrift,
with sails blown from the yards and top-gallant masts gone. At
3.30 P.M., hurricane at its height, blowing so terrifically hard that
it was impossible to stand on deck without holding on. At this
time our inshore bower chain parted, our sails were all blown from
the yards, and the top-gallant mast went with the foretopmast.
When the bower chain parted we swung out stern on to the gale,
and held on for a few minutes, when in a tremendous burst of
wind our stern chain parted, and away we drove across the river
before wind and tide, at a frightful rate, smashing into several
ships on our way. Finally, we were brought to a standstill on the
opposite ride of the river, and became a target for one half of the
ships in Calcutta. One wooden ship driving up struck upon our
starboard quarter, walking right through the upper part of our
stern, and raising the poop deck. Three or four ships were
constantly pitching into our main rigging, being all fast together,
and smashing and tearing away at everything thenceforward. At
4.30 P.M. two iron ships and one wooden one drove right into us
abaft the forerigging, carrying away chain plates and rails. One

of their bowsprits struck the foremast, and, with a fearful crash, the foremast fell over the port side, almost burying a small vessel that was fast to us. The rigging of the foremast was totally gone. Some time before the mast went it broke 'tween-decks, tearing up the main deck, and breaking two beams. 5.30 P.M., wind abating very fast, and barometer rising, with fine weather. Ship lying almost a helpless wreck.

" Hard usage enough for $\frac{3}{8}$-inch steel plates," says another contemporary account,[1] " endured, nevertheless, in a way sufficient to place steel in the foremost position as a material for shipbuilding. At one point near the forerigging the *Clytemnestra* was struck fairly end on by the *Glenroy*, a ship of 1,000 tons burden. The plates were beaten in, but not fractured. Forward, the continual hammering of several large vessels beat the bulwarks level with the deck ; the plates forming them were, nevertheless, so tenacious that they were prized back to their original position, and made to do duty again without the aid of the riveter. In another part of the bulwarks a plate had been partially knocked out, and, catching against the side of the other vessel, was rolled up as perfectly as a sheet of paper could be. In the stern between the upper deck and the poop several plates were driven in by repeated blows from a heavy wooden ship. These and the angle irons were twisted into a hundred fantastic forms, in some cases doubled and redoubled, and in no case was there a crack or fracture that indicated any brittleness in the metal. Notwithstanding all the hard usage she received, the *Clytemnestra* did not make one drop of water, and afterwards made a voyage to the Mauritius, back to Calcutta, and thence home, with no further repairs to her hull than the ship's carpenter was able to effect." The account we quote concludes by stating that the vessel was then (October, 1865) lying in the Liverpool docks, " bearing her honourable

---

[1] *Engineer.*

scars as so many awards of merit and testimonies to the accuracy of all that can be said in favour of Bessemer plates."

But all these severe trials made little or no impression on incredulous shipbuilders. True, the *Altcar*, a similar vessel in size to the *Clytemnestra*, was also built of steel in 1864 ; and in the following year six large ships were built of Bessemer steel, having an aggregate tonnage of 5,342 tons. In his paper read before the British Association at Birmingham in 1865, Sir Henry Bessemer stated that the firm in question had up to then constructed no less than 31,510 tons of shipping, wholly or partly of steel ; of these, thirty-eight vessels were propelled by steam, and, besides that, the principal masts and spars of eighteen sailing ships had been made by them wholly of steel. He also showed that steel vessels could carry 25 per cent. more of cargo than iron ones, on account of the less weight required of the superior metal—a result now generally admitted. Unfortunately the enterprising firm that Sir Henry Bessemer induced to build these ships suddenly got into financial difficulties at the time when two fine steel vessels, each of 1,622 tons, were lying unfinished in the yard. Sir Henry was appealed to for money to finish them on the ground that it was of immense importance to the future of steel shipbuilding that these vessels should be finished. In response to this appeal he put down £10,000 for their completion ; but soon after their completion the firm went into liquidation, and for the next ten years steel shipbuilding was almost unheard of. Its revival will be recorded in a subsequent page.

Meanwhile Sir Henry's fertile brain and irrepressible ardour were busily employed in devising other uses for his steel. About the middle of January, 1864, there appeared in the *Times* an account of some experiments made with Bessemer steel spherical shots fired from a smooth-bore gun against a $5\frac{1}{2}$ inch armour-plate ; and Sir Henry, while remarking that their destructive effects as compared with projectiles made of cast or wrought iron

entirely confirmed the views he had so long advocated in vain,
gave the following interesting particulars :—

"It is now just three years since I obtained a patent for producing cast steel spherical shot by a peculiar arrangement of the
rolling mill, by means of which spherical steel shots may be made
with rapidity and correctness. I also exhibited a spherical steel
cannon ball at the International Exhibition of 1862 for the purpose
of giving further publicity to my views on this important question ;
but it is only after this lapse of time that a trial is made of them
in England, although a delay of ten or twelve days, and an
expenditure of 50*l.*, would have given as full a proof of their
efficiency three years ago as we have to day. Meanwhile, however, many hundred thousand pounds have been expended in
building iron-plated ships, which these long-neglected steel projectiles will riddle as easily as the cast-iron shot found its way
through the wooden walls of our old men-of-war. It is marvellous how the advantages of using such a material for projectiles
did not force itself on the attention of every practical artillerist,
irrespective of any efforts on my part, for there is scarcely a
schoolboy to be found who does not know that a snowball flung
with great force is perfectly harmless, while a stone or other solid
substance of equal weight would inflict a severe injury, simply
because the snowball will fall to pieces on striking the object,
while the stone will remain entire, and consequently administer
the whole force with which it is thrown. Now, the way in which
a cast-iron shot is broken, and scattered in a shower of small
fragments on striking an armour-plate, bears a very strong analogy
to the snowball in the case supposed.

"It is not less remarkable that while our firm have manufactured
at Sheffield some 150 pieces of Bessemer steel ordnance for
foreign service, guns made of this material are still untried by
our government, although it is well known that the strength
of this metal is double that of ordinary iron, while such is the

facility of production that a solid steel gun block of twenty tons in weight can be produced from cast fluid iron in the short space of twenty minutes, the homogeneous mass being entire and free from weld or joint.

" Our armour-plate system has certainly received a severe shock, and it behoves us now to see how far it is possible to increase the resisting power of ships so as to keep pace with the advances made by steel shot. On the 12th of December, the fine ship *Minotaur* was launched from the yard of her builders at Blackwall. She was all that excellent workmanship and the best iron could make her, but still she was only iron.

" It has been stated that the hull of the vessel weighs 6,000 tons, and her $4\frac{1}{2}$-inch armour 1850 tons. Now, had the hull of this vessel been built of a material possessing double the strength of ordinary iron her weight might have been reduced to 3,000 tons ; but suppose that, while we admit a double strength of material, we only reduce the weight by one third, this would give 4,000 tons of steel for the hull. Now with this reduction in the weight of the hull, we may employ 9-inch armour plates in lieu of the $4\frac{1}{2}$-inch armour plates now employed. It must be borne in mind that the resistance offered by the armour plate is equal to the square of its thickness : hence a vessel constructed in the manner proposed would bear a blow of four times the force that the present structure is calculated to withstand.

" Thousands of Bessemer steel projectiles are now being made for Russia, and from undoubted sources I learn that other orders for steel shots have been given to the extent of 120,000*l.* in value. Have we a single ship afloat that can keep out these simple round steel shots fired from a common smooth-bore gun, if ever directed against us ? This is a grave question, and demands a speedy answer."

The practical answer to this grave question was given by the British Government twelve years afterwards !

CHAPTER V.

" When great inventions or discoveries in the arts and sciences either abridge or supersede labour—when they create new products, or interfere with those already existing—the advance which is thus made involves not only a grand and irrevocable fact in the progress of truth, but it is a step in the social march which can never be retraced."—BREWSTER.

THE year 1866 may be said to mark a turning point in the inventive work of Sir Henry Bessemer. In each of the previous thirty years he had designed at least one invention, and in some years as many as ten. 1866 was the first year since 1838 in which no new patent was taken out by him; his greatest invention—the Bessemer process—was now bringing him in an income of 100,000*l.* a year; and when 1866 showed a blank in his record of inventions, it might reasonably be assumed that henceforth he would rest from his labours. But next year his name again appeared in the list of new patentees, and he continued to take out patents for new inventions without interruption for the next ten years. Yet the period of his greatest activity was the eventful years which he devoted to perfecting his process of steel making. Having satisfied himself that the principle of that process was sound and practicable, he took out a fresh patent for each new improvement or substantial alteration, in order to secure his rights intact. Thus he took out fourteen patents in 1855, ten in 1856, and six in 1857, making thirty patents within three years, not to mention his numerous foreign ones. As showing how necessary it was to patent every important alteration or improvement, he has himself explained that when great excitement was caused by the reading of his first description of the Bessemer process at Cheltenham in 1856, many persons seemed to covet a share in an invention that seemed to promise so much. Consequently there was a general rush to the patent office, each one intent on securing his supposed improvement. It was thought scarcely possible that the original inventor should at the very outset have secured in his patents all that was necessary to the success of so entirely novel a system; he must surely have overlooked or forgotten something; perhaps even left out all mention of some ordinary appliance too well understood to really need mentioning; so in the jostle and hurry to secure something, any point on which a future claim could be reared was at once patented. Some of these claimants even repatented portions of his own patents, while others patented things in daily use in the hope that they might be considered new when added to the products of the new process. But all in vain!

The Commissioners of the Great Exhibition of 1862 stated that during the previous eleven years, from May, 1851, to May 1862, no less than 177 applications for patents had been made for improvements in the manufacture of steel, and 127 of these had been actually sealed. Yet out of these 127 patents, they add, there is only one which has brought about any striking change in the mode of the production of steel, or which has been attended with any really practical or commercial results, namely, the Bessemer process.

It need scarcely be said that Sir Henry Bessemer is a believer in patents; but to his varied experience in the introduction of new inventions, another singular fact has to be added. " I do not know," he says, " a single instance of an invention having been published and given freely to the world, and being taken up by any manufacturer at all. I have myself proposed to manufacturers many things which I was convinced were of use, but did not feel disposed to manufacture or even to patent. I do not know of one instance in which my suggestions have been tried; but had I patented and spent a sum over a certain invention, and saw no means of recouping myself except by forcing, as it were, some manufacturer to take it up, I should have gone from one to the other and represented its advantages, and I should have found some one who would have taken it up on the offer of some advantage from me, and who would have seen his capital recouped, by the fact that no other manufacturer could have it quite on the same terms for the next year or two. Then the invention becomes at once introduced, and the public admits its value; and other manufacturers, like a flock of sheep, come in. But the difficulty is to get the first man to move. The first man might say: 'Oh, my machinery cost me a great deal of money; I have my regular trade, and this new scheme is sure to be more trouble to me in the first instance; and when everybody asks for it, every other manufacturer will be in a condition to supply it; so it is not worth my while.' I believe inventions which are at first free gifts are apt to come to nothing."

Altogether Sir Henry has taken out 120 patents for inventions, and it is calculated that he has paid to the Patent Office 10,000*l.* The specifications of these patents fill two bulky volumes, and the drawings illustrating them, every one of which was executed by himself, fill seven large volumes.

Always eager to design something new, he applied himself in later years chiefly to the solution of problems which were more likely to benefit the public than to increase his own princely fortune. He continued therefore to make new inventions and to demonstrate their practicability by numerous and costly experiments, but he did not push their application with the energy and perseverance which were necessary to bring such a self-evident boon as his converter into general use.

The success of his converter naturally increased his interest in metallurgy, which ever after had a foremost place in his mind. Accordingly in 1868 he patented another invention which he had constructed with great labour, and which he expected would be a benefit to the iron and steel trades. In principle it was another ingenious application of the oxygen of the air to increase combustion so as to give excessively high temperature applicable on a large scale to industrial purposes. The electric light and the oxy-hydrogen blow-pipe were then used in laboratories to give extreme heat, but they were not applicable to great manufactures. Our ordinary source of heat, coal, gave a maximum temperature which was insufficient to fuse many substances, however long they were exposed to it. Sir Henry determined to supply this deficiency. It is a known law of nature that all gaseous bodies increase in volume with increase of temperature. An addition of 480 degrees Fahrenheit so expands one cubic foot of gaseous matter that it occupies two cubic feet, and so on. It appeared to Sir Henry that if the same amount of heat that existed in a certain space were condensed into a smaller space of one half the size, the temperature would be doubled. Accordingly he designed a furnace inclosed in a strong iron case, like a steam boiler. Into this furnace air was forced at a pressure of 20 lb. to the square inch; and as its only means of escape was a small hole $1\frac{1}{2}$ inch in diameter, the internal pressure was kept at 15 or 16 lbs. above that of the external air. In other words, the air inside the furnace was twice as dense as that of the atmosphere. By this means all the heated gaseous products of combustion were compressed into one half the space they would have occupied in an ordinary furnace, and there was a corresponding increase of temperature. In this furnace a piece of cold bar iron 1 foot long and 2 inches square, weighing $13\frac{1}{2}$ lb., was melted in $5\frac{1}{2}$ minutes to the fluidity of water. Any other furnace would have taken 2 or 3 hours to

effect this change. In another experiment 3 cwt. of malleable
iron put in cold were fused in 15 minutes. With these results
before him the inventor stated his conviction that nearly every
substance known to man might be fused on this system. But
would not the furnace itself be the first to be fused? This was
the great difficulty that he had to meet; and after much consideration and many experiments, he succeeded in making a
furnace lining which, at this high temperature, was less rapidly
destroyed than the linings of ordinary furnaces. This new lining
was of ganister, considered the most refractory material for such
purposes. It was 3 inches in thickness, and was rammed close
to the iron casing. Outside of this iron casing was placed
another, and between the two cold water was made to circulate so
freely that it kept the inner iron casing always cool. By this
arrangement the part of the inner ganister lining in contact with
the first iron case was kept black cold, the part mid-way between
the cool iron and the internal fire was blood-red, while the face
of the lining in contact with the fire was as white and brilliant as
the sun. The inventor stated that at first the increase of heat
caused some of the face of the ganister lining to melt off, but it
would always come to a point where the cooling influence of the
external wall of water would prevent further melting away, and
here the lining would become permanent.

It was at the time that Sir Henry Bessemer was thus employed
that a Royal Commission was inquiring into the probable duration
of England's coal supplies; and one of the questions that came
before it was the depth at which coal could be worked. The
question is still a perplexing one, and is likely to be so for many
years to come. The Royal Commission reported that the rate of
increase of the temperature of the strata in the coal districts of
England is in general about one degree of Fahrenheit for every
sixty feet of depth; that the temperature of the strata operates as
an impediment to deep working by heating the air circulating

through the passages of the mine ; and that the depth at which the temperature of the earth would amount to blood heat, or 98 degrees, is about 3,400 feet, or two-thirds of a mile Writing in 1869 with reference to coal-mining, Sir Henry said : " I am at the present time busily engaged in investigating the action of combustion under an excessive pressure in furnaces where the flame is bottled up (so to speak) like steam in a steam boiler, by which means the heat is intensified in the ratio of the pressure employed, so that the most refractory substances known to man may be fused or dissipated in vapour with the same quickness and facility with which our most easily fusible substances are melted. In one modification of this furnace the workmen operate in a large room where the pressure of the atmosphere is greater than it would be at a depth of ten miles below the surface of the earth and where the temperature under ordinary circumstances would be such that no attendant of a Turkish bath could endure it for a single hour. Yet these men and the furnace they attend may by a simple arrangement of the apparatus be supplied with thousands of cubic feet of air per minute as cool, or, if necessary, much cooler, than the surrounding atmosphere. The coal miner inclosed all day between black masses of coal above and around him requires a powerful light to see what he is doing —a light that never fails, never goes out, never requires trimming, and, above all, a light that effectually prevents the mixture of air and gas that ever pervades all coal mines from encountering the flame and becoming ignited. Now these are precisely the conditions obtained by combustion under pressure, which offers to the miner a source of brilliant light, while insensible to the inflammable air of the mine. As a simple illustration of the fact, let us suppose a simple box, a little larger than a policeman's lantern, having a thin plate glass or a bull's eye on one side of it ; in the lower part is a common gas burner supplied by a pipe from a gasometer above ground

The supply of air to support combustion is arranged in a similar manner, and supplied under pressure above ground. A small aperture is made in the top of the lantern for the escape of the products of combustion. Now if air and gas are supplied to this light under a pressure of, say, one pound per square inch, the light would be brilliant, and the escape from the orifice at this pressure (or even far less) would prevent the possibility of any external gases entering and becoming ignited. In this way every gallery in a mine may be lighted like a workshop, to the great comfort and cheerfulness of those whose whole lives are spent in the cheerless gloom of those dangerous workings."

His high pressure furnace has never been put into practical operation in the iron trade ; but the ingenious means by which he preserved its lining from fusing was experimentally applied in 1881, in a modified form, to preserve the lining of some blast furnaces in Germany, with satisfactory results.

At the same time that Sir Henry was experimenting with his model high pressure furnace he was directing his attention to the invention of a direct method of converting iron ore into steel —an operation that had never been successfully done before on a large scale. He constructed a novel apparatus for this purpose, but had only made three or four imperfect experiments with it when his health failed. He had become so absorbed in his experiments for perfecting his high pressure furnace, and was in such haste to get in his patents, that the prolonged strain undermined his health ; his medical adviser forbade any further experiments with his new apparatus, and he found it necessary to adhere to that advice. The completion of his new process for the conversion of iron ore into steel was thus abandoned for a time, and ere the subject was resumed his most distinguished contemporary announced an efficient process which he had invented with the same object.

Another invention met with a different fate. Its history affords another illustration of the saying that it is one thing to make a good invention and another thing to bring it into general use.

In 1871 Sir Henry Bessemer was selected to succeed the Duke of Devonshire as president of the Iron and Steel Institute ; and in the course of his inaugural address he made the following statement : "Among the most important improvements lately effected in the manufacture of steel is the development, by Sir Joseph Whitworth, of the system of casting under hydraulic pressure. The casting of large masses of steel free from air bubbles has long been a source of difficulty, owing chiefly to the fact that at the extremely high temperature of molten steel a certain quantity of oxygen is absorbed and retained by the metal, so long as this high temperature is kept up, but it cannot keep this oxygen in combination when the metal is cooled down to the point at which it commences to solidify ; hence, when the fluid metal is received in a cold mould, large volumes of gas are given off, some of which becomes entangled in the solidifying mass, and is there retained, forming numerous cells or honeycombs. A similar result is met with in finery iron, when it is ' overblown ; ' carbonic oxide gas is liberated in abundance during the solidification of the plate metal and gives rise to the peculiar cellular structure so well known. Another defect inherent in steel castings owes its origin to the crystalline structure assumed by the metal in the act of solidification. So long as the metal retains undisturbed the original crystals formed by casting, the mass is only feebly coherent, its tensile strength is less than half that to which it rises when hammered or rolled. It will bend only a few degrees from the straight line without fracture, while its power of elongation is also extremely limited ; but if considerable pressure be applied while the steel is passing from the liquid to the solid state, the crystals, which would otherwise become almost independent structures, are united or welded together so perfectly at this high temperature and in their almost plastic state as to develop the most perfect cohesion of all the parts of the mass, probably more perfect than any subsequent operation of hammering could effect. In a patent which I obtained in 1856 I described a method of casting steel under hydraulic pressure in iron moulds—a cold wrought iron plunger being forced into the semi-fluid steel at one end of the mould through the agency of the hydraulic pressure applied to its opposite end. About the same period I had observed that in those cases where fluids gave off gaseous matters under ordinary atmospheric pressure they were prevented from doing so by increasing the pressure on their surfaces. A familiar example of this action is seen the instant we release the gaseous pressure by the removal of the cork from a bottle of champagne, and it occurred to me that if I subjected the fluid steel to additional atmospheric pressure, the boiling of the metal in the mould would be prevented. Thus arose the first idea of casting under the pressure of air or gases pumped into a close chamber of great strength, in which the mould and casting were inclosed; but owing to numerous engagements these inventions were left in abeyance until attention was again called to the subject a few years since by Sir Joseph Whitworth, who finding great difficulty in making steel castings free from air bubbles and of sufficient cohesive strength for the manufacture of his guns and projectiles, hit upon the idea of subjecting the metal, while still fluid, to the action of a hydraulic plunger forced into the mould. His experiments in connection with this system of casting have been most successful. Indeed, I can bear witness to the extreme soundness of several large cylindrical masses, turned and bored, which were shown to me at his works, in neither of which the most minute flaw or bubble hole was visible. In justice to Sir Joseph Whitworth I feel bound to say that I have no doubt whatever but that he was wholly unaware of the existence of my previous invention at the time he brought forward his system of casting under hydraulic pressure, by which the material

now known as 'Whitworth steel' is produced. Certain it is
that we owe to him the development and first practical application
of this system of casting steel."

Sir Henry's patent is on record as evidence of his originality, but when he delivered the above address he had
evidently forgotten that he was the inventor of another and a
simpler process which has since been proved to exclude or
occlude the gases from steel nearly as effectually as the costly
process of compression. In 1863 he patented a mechanical
agitator or stirrer intended to stir the molten steel in the ladle,
as a painter stirs paint in a pot, in order to mix the various
ingredients of which it is composed; but somehow or other this
plan was allowed to lie dormant. It was never worked during
the life time of the patent, but after the latter had expired, Mr.
W. D. Allen, in 1878, tried the use of this mechanical agitator
at the Sheffield works of the Henry Bessemer Company. It
turned out to be a great improvement. As the operation is
described by Mr. Allen, the mixing or stirring takes place in
the ladle immediately before casting. The agitator is simply
an iron rod about $1\frac{1}{2}$-inch diameter, one end of which has a
long aperture made in it, through which is inserted the blade
or plate of iron, about 2 feet long, 4 inches to 5 inches wide,
and about $\frac{3}{8}$ inch thick. The blade thus inserted, is twisted
at each end, so as to give it somewhat the form of a screw
propeller blade. The rod and blade are coated over with loam
or ganister, which has to be thoroughly dried, blacked, and
carefully prepared. The other end of the rod is attached to the
apparatus which drives it. The ladle of steel, immediately it is
turned out of the vessel, is brought beneath the agitator, and
raised by a hydraulic crane, immersing the blade and a portion of
the rod in the steel. Rotatory motion of about 100 turns per
minute is then given to it, the ladle being lowered and raised
again during the operation to ensure all portions of the steel

being operated upon. When the stirring is deemed sufficient, the ladle is lowered clear of the agitator, and the casting is proceeded with in the usual way. The evolution of gas is very manifest during the stirring process, particularly at its commencement, when gas is seen to force its way through the slag covering in large quantities, frequently with a considerable roar, and the ebullition of the steel is sometimes so violent as to cause the metal to rise over the sides of the ladle, and the stirring to be moderated. "The simple fact is," says Mr. Allen, "that the violent frothing so frequently seen in the moulds whilst casting is got rid of in the ladle by the aid of the agitator, and sound castings free of all honeycombing and uniform throughout are now made with perfect regularity and certainty by the Bessemer process. Every ingot formed from the largest charges is now found by analysis to be perfectly uniform in temper and quality, while the thoroughly homogeneous quality of every part of the same ingot is evinced by its behaviour under the hammer or in the rolls, as well as in hardening and tempering. The stirring operation is found to be very simple in practice, causing no delay or inconvenience of any kind, and costing almost *nil.*"

It thus appears that of the three methods now in use for the production of steel without blow-holes Sir Henry was the first inventor, and none of them ever yielded him any direct benefit The application of silica (sand), the origin of which is narrated in the previous chapter, has been most successfully used in France and Sweden.

Sir Henry Bessemer in 1869 entered upon a new field of invention. Knowing the inexpressible discomfort of sea-sickness, he determined to construct a saloon for a passenger vessel, in which the vibrating motion that is supposed to cause sickness should, if possible, be prevented or minimised. For this purpose he designed a new application of the hydraulic principle that he had successfully utilised in other inventions. Previous success

I

inspired confidence in the use of this power. During the Crimean war, steel cannon balls were made in London by being turned in a lathe, and the time required to finish a single shot in this way was two days. By the application of hydraulic power, for the production of three eighty-four pound shot at one time, Sir Henry made the rough mass of steel into perfect spheres, eight inches in diameter, in four minutes. The same powerful agent was also successfully applied by him to the production of endless steel railway wheel tires; but his greatest success was in its application to the movement of the large converters used in the manufacture of steel. By hydraulic power the movement of these vessels, weighing from ten to twenty tons, was placed under the absolute control of a boy standing sixty feet away from them. By turning a handle the boy could so manipulate and regulate the motion of a converter that at will he could pour out a fine stream, or let fall a seething flood of incandescent metal into the receiver. This apparatus was capable of altering the position of the vessel to the fractional part of an inch, or to move it round a circle, although the fluid contents, weighing five tons, were constantly altering their balance. It was by a similar application of the same power that he hoped to counteract the wave motion of a vessel at sea, that was supposed to be the cause of sea sickness. He proposed to place in the centre of a large passenger steamer a swinging saloon, which would be so governed by powerful hydraulic apparatus connected with the under side of the flooring, that, as the vessel rolled on either side, the pressure or resistance afforded by water would be instantly brought into play and utilised in checking the rolling motion. A model of the novel apparatus was erected on his estate at Denmark Hill, and it was found to realise his expectations. A small saloon ten feet long was controlled by a pair of sensitive equilibrium valves connected by a hand lever. By means of this hand lever a steersman, who watched the slightest indication of rolling by observations on a spirit level, instantly suppressed the slightest tendency of the saloon to follow the motion of the ship. The working of this model was inspected by the most eminent engineers in England. Distinguished foreigners also visited Denmark Hill to see it. All admired it. A company was soon formed to work it in a sea-going vessel; and by order of this company Sir E. J. Reed designed a vessel, 350 feet long, specially adapted for the reception of a swinging saloon 70 feet long by 30 feet wide, and 10 feet high The vessel was launched on September 24, 1874, and was named the *Bessemer.* The swinging saloon, which was then described as the greatest wonder of its kind since the hanging gardens of Babylon, weighed 180 tons. The trial trip of this unique steamer, intended for the Channel service, took place in May, 1875. The result of the trial was awaited with eager curiosity, for since the first announcement of its construction the question had been keenly debated in the public press, whether sea sickness could be averted by the most successful mechanism. Not a few eminent medical men maintained that it could not; but only the test of experience could settle the point. The trial trips of the vessel between Dover and Calais disappointed expectation. The weather was fine and the working of the saloon was little needed; but the impression gained ground that however nicely the swinging saloon could work, it would not have the desired effect of averting that dreaded malady which is as inexplicable in its nature as it is capricious in the choice of its victims. It was found, too, that Calais harbour was too small for such a vessel. In two out of three voyages serious damage was done to the pier, and these accidents involved the company in heavy expenses. In these circumstances the costly experiment, towards which Sir Henry had contributed 25,000*l.*, was abandoned.

An invention of a very different kind next occupied his attention. It was the construction of a novel telescope at a cost exceeding 20,000*l.* Hitherto in large telescopes a metal speculum had to be used, because it had been found impracticable to grind glass to the required curvature. The first thing he did, therefore, was to invent a machine for supplying this desideratum. This was no easy task, on the contrary it was enough to appal any ordinary mind; but its success promised great results. Like all the great inventions in the manufacture of steel, the reflecting telescope was a British invention, and up to the present time all the most important improvements in its construction have been made in the United Kingdom. As in the steel trade, too, the greatest inventions have not been made by professional metallurgists, so the scientific world has been more indebted to amateurs than to regular opticians for improvements in the telescope. A few particulars as to the labours of his greatest predecessors in the construction of gigantic telescopes will best indicate the magnitude of the work which Sir Henry Bessemer commenced in his sixty-fifth year. Just a century had elapsed since Sir William Herschel, then an organist, had made his first reflecting telescope with his own hands at the age of thirty-six, and his labours were considered for many years afterwards to have brought the invention to perfection. *Apropos*, one of his biographers says that " to construct good telescopes is itself no easy art, especially if they are to be carried to a size and perfection previously unattained. To ascertain the best composition, whether of glass or metal, to melt and to cast it in the right way, is one branch of the art which may be called the chemical part. To fashion the lens or mirror correctly by grinding, and to fit it for optical use by giving it an exquisite polish, is a second step requiring a peculiar mechanical skill and perseverance. To mount the telescope effectively is another and entirely different problem in mechanics ; and all these the amateur astronomer must be prepared to accomplish with his own hands, unless he command the services of practical opticians to an extent rarely to be bought." Sir William Herschel has stated that at Bath he constructed 200 specula of seven feet focus, 150 of ten

feet, and about 80 of twenty feet; so that from this great number
he could select those having the most perfect figure. Afterwards
he contrived a method of obtaining accurately the parabolic form,
but he kept his method a secret. The Earl of Rosse, who
subsequently carried the construction of the speculum a step
further, has given an account of his method of casting and grinding.
He not only constructed in 1844 a telescope larger than any that
had been made before, but adapted machinery, driven by steam,
to the grinding and polishing of the mirror, and by this invention
the largest speculum could be made nearly as quickly and accurately as the smallest. Lord Rosse began his labours which ended
in this achievement in 1828, and they were continued for sixteen
years with an enthusiastic perseverance and mechanical skill rarely
displayed in one of his rank and wealth. He had to instruct his
assistants in his own workshops, and had to design and manufacture
the steam engine which worked his own polisher. He made many
experiments before he could overcome the extreme difficulty of
procuring large castings of so excessively brittle a material as
speculum metal; but at last he succeeded in making at a single
casting immense mirrors, which, when ground and polished, were
truly parabolic. The speculum of his largest telescope was polished
in six hours; it weighed four tons, and its surface measured twice
that of the largest ever made by Sir William Herschel. Apart
from the speculum, the construction of these great telescopes
entailed many difficulties in detail, which would occupy too much
space even to mention here.

   Undaunted by the great labours in which the construction of
an improved telescope had involved these men, Sir Henry Bessemer
determined to spend the closing years of his life in effecting a
revolution in the art of telescopic construction. In 1875 he
invented a lathe capable of giving to large pieces of rough or
plain glass the general form or curvature required for telescopes
and other optical instruments and reflectors. The lathe designed

for this purpose had two " face plates," each larger than the largest lenses, and on the perfectly flat surface of these plates concentric annular channels were indented ; so that when a circular piece of glass was placed against one of the plates, a vacuum could be formed in the annular channels by an air-pump, and the atmospheric pressure on the exterior surface of the glass caused the latter to adhere firmly to the face of the plate. Thus fixed, the inventor says the glass does not become bent or distorted as it would be by the old process of beating it on a soft adhesive or yielding material ; it is so truly supported as not to be easily fractured even by repeated heavy blows. Attached to this is an apparatus which guides a tool with a diamond set in it ; and this diamond, applied to the glass firmly fixed on one " face-plate," cuts a concave lens ; and, applied to the glass set against the other " face-plate," it cuts a convex lens.

Having thus invented a machine for cutting a larger glass speculum than had ever been made before, Sir Henry commenced in 1877 the construction of a telescope designed to magnify 5,000 times, and possessing many novel features. For several years he had cherished the desire to construct such an instrument as would be at least equal in power to any previously constructed ; and it was his special wish that this instrument, notwithstanding its great magnitude, should be placed in a commodious and comfortable observatory, instead of standing in the open air exposed to all changes of weather He further desired that the instrument should be capable of being directed to any part of the heavens at will, without either waiting for the earth's motion—as in the case of Lord Rosse's great telescope at Parsonstown—or for the assistance of any one but himself, so that at any time of the night he could use it without climbing up long ladders, or lying on his back in an uncomfortable and constrained position. With this view he designed a new form of telescope, at which the observer could sit or stand in an upright position at the centre of the floor of the observatory, looking straight before him into the eye-piece, which he placed five and a half feet above the floor.

The construction of such a powerful telescope, possessing this great advantage, marks an epoch in the history of the instrument, the effectiveness of a large telescope being enormously increased when the observer is thus enabled to pursue his work in comfort. The device for securing this result is worked by hydraulic power. The observing room, with its floor, windows, and dome, revolve and keep pace automatically with every motion of the telescope, notwithstanding that the latter is wholly detached from the moving parts of the building, and stands firmly in the centre on a massive masonry and concrete foundation, from which the upper end of the telescope reaches an altitude of forty-five feet when the tube is vertical.

In order to make its range of vision as great as its mechanical arrangements are comfortable, Sir Henry devoted himself to the study of Herschel's works on optics, and brought into use some principles that had lain dormant there for many years ; while, in order to complete his instrument, he constructed special machinery for the purpose of finishing a four-feet mirror made of silvered glass, instead of the metal used in previous great telescopes, such a mirror possessing far greater power of gathering light, as wel as being far more manageable. Novel and ingenious as the design appeared, eminent authorities did not hesitate to say of it, before completion, that not only would this telescope be more effective than any yet made, but it would be the forerunner of others yet more powerful, while of the mechanism which he invented to give the true parabolic form to the reflector with a degree of mathematical precision never yet attained in large specula, it was anticipated that after his success in this as in the turning of the glass to shape with a diamond, "a perfect glass reflector would in future not cost as many shillings as it previously cost pounds ; and what is of still greater importance, his system of mounting comparatively thin plate-glass on a marble backing holds out the prospect of our future telescopes being made at least four times as large as his own one."

The Bessemer telescope stands on the Bessemer estate at Denmark Hill, London. That estate, as described by an appreciative writer,[1] is some 30 or 40 acres in extent, and is on the slope of a hill looking towards the Crystal Palace. " From the terrace of Sir Henry Bessemer's house the distant view is singularly pleasing and essentially English in its unhidden sylvan character. The foreground of the picture is almost entirely manufactured. Its natural slope towards the valley is perhaps generally maintained, but hills have grown where there were depressions, rocks have sprouted forth where there was nothing but gravel, and a lake with its feeder and outlet forms part of the purely artificial river system. Beyond the superbly terraced lawn is a sweet bit of real nature, a meadow in which the tall rye grass, buttercups and ox-eyed daisies wave at every puff of wind, making the rich carpet red, yellow, and white by turns. At the lower side of this flat meadow is a bowling green, level as a billiard table, and separated from it by a rim of ground ivy, next to which comes a triumph of landscape gardening—a great clump of rhododendrons on the rocky shore of the picturesque lake, ' Nowhere,' Sir Henry Bessemer tells his guests, ' more than 18 inches deep, an excellent depth for summer boating and winter skating, and most convenient if a child falls out of the boat or the ice gives way.' Around the lake are thick shrubberies, in which hawthorn, jasmine, and honeysuckle contend with the lilac, laburnum, laurel, and innumerable rhododendrons. Presently the host vanishes in a bush as if by pantomimic trick, and in a moment reappears and leads the way into a cavern filled with a magnificent collection of ferns, heated to the precise temperature, and lighted by a skilful combination of toplights and mirrors. At one extremity of this cave,

[1] *The World.*

lined with rocks made of brick and cement, is a waterfall pouring
over a glass wall ; at the other a snug little smoking-room looking
over the lake, with all necessary refreshments hidden behind a
rock apparently as massive as a cheesewring, and of nearly the
same outline."

In literature Sir Henry Bessemer has done very little except to
write an occasional paper explaining his own inventions ; but his
few contributions on subjects of more general interest are of
permanent value. Some discussion having taken place in the
public journals in 1878 on the question, What is a billion? Sir
Henry gave the following exposition of it :

"It would be curious to know how many have brought fully
home to their consciousness the significance of that little word
billion. Its arithmetical symbol is simple and without much
pretension. There are no large figures, just a modest 1, followed
by a dozen ciphers, and that is all. Let us briefly take a glance
at it as a measure of time, distance, and weight. As a measure of
time I would take one second as the unit, and carry myself in
thought through the lapse of ages back to the first day of the year
one of our era, remembering that in all these years we have 365
days, and in every day just 86,400 seconds of time. Hence in
returning in thought back again to this year of grace 1878, one
might have supposed that a billion of seconds had long since
elapsed. But this is not so. We have not even passed one
sixteenth of that number in all these eventful years, for it takes
just 31,687 years, 17 days, 22 hours, 45 minutes, and 5 seconds
to constitute a billion seconds of time.

"It is no easy matter to bring under the cognisance of the
human eye a billion of objects of any kind. Let us try in
imagination to arrange this number for inspection ; and for this
purpose I would select a sovereign as a familiar object. Let us
put one on the ground and pile upon it as many as will reach 20
feet in height. Then let us place numbers of similar columns in

close contact, forming a straight line and making a sort of wall 20 feet high, showing only the thin edges of the coin. Imagine two such walls running parallel to each other, and forming, as it were, a long street. We must then keep on extending these walls for miles, nay, hundreds of miles, and still we shall be far short of the required number. It is not till we have extended our imaginary street to a distance of 2,386½ miles that we shall have presented for inspection our billion of coins. In lieu of this arrangement we may place them flat upon the ground, forming one continuous line like a golden chain, with every link in close contact. But to do this we must pass over land and sea, mountain and valley, desert and plain, crossing the equator and returning round the southern hemisphere, through the trackless ocean, retrace our way again across the equator, then still on and on till we again arrive at our starting-point; and when we have thus passed a golden chain round the whole bulk of the earth we shall be but at the beginning of our task. We must drag this imaginary chain no less than 763 times round the globe. If, however, we can imagine all these rows of links laid closely side by side, and every one in contact with its neighbour, we shall have formed a golden band round the globe 56 feet 6 inches wide ; and this will represent our one billion of coins. Such a chain, if laid in a straight line, would reach a fraction over 18,328, 445 miles. The weight of it, if estimated at a ¼ oz. each sovereign, would be 6,975,447 tons, which would require for their transport no less than 2,325 ships, each with a full cargo of 3,000 tons. Even then there would be a residue of 447 tons, representing 64,081,920 sovereigns.

" For a measure of height, let us take a much smaller unit as our measuring rod. The thin sheets of paper on which the *Times* is printed, if laid out flat and firmly pressed together as in a well-bound book, would represent a measure of about $\frac{1}{333}$rd of an inch in thickness. Let us see how high a dense pile, formed by a billion of these thin paper leaves, would reach. We must in imagination pile them vertically upward, by degrees reaching to the height of our tallest spires; and passing this the pile must grow higher, topping the Alps and Andes, and the highest peaks of the Himalayas; and shooting up from thence through the clear clouds, pass beyond the confines of our attenuated atmosphere, and leap up into the blue ether with which the universe is filled, standing proudly up far beyond the reach of all terrestrial things. Still pile on your thousands and millions of thin leaves, for we are only beginning to reach the mighty mass. Add millions on millions of sheets, and thousands of millions on these, and still the number will lack its due amount. Let us pause to look at the neat ploughed edges of the book before us. See how closely lie those thin flakes of paper; how many there are in the mere width of a span; and then turn our eyes in imagination upwards to see the mighty column of accumulated sheets. It now contains its appointed number, and our one billion sheets of the *Times* super-imposed upon each other and pressed into a compact mass has reached an altitude of 47,348 miles."

Another specimen is still more interesting and beautiful. In the form of a letter to the *Times* he wrote the following in the spring of 1882:

" The Easter holidays have come round once more, and our boys, with their bright, beaming faces, full of mirth and cheerful-ness, have been flocking home from school to dear old smoky London, all unmindful of its murky atmosphere, and intent only on the many wondrous sights they hope to see. I had just filled some loose sheets with calculations which I had been making, with a view to afford some amusement to my grandsons on their return, when, looking up from my task, I noticed a stream of small, bright objects flitting by. The sharp east wind was breaking up the large seed pods on the great Occidental plane tree near my study window, and its taper seeds, with their beautiful little gold-coloured parachutes, were being wafted far away, falling into little chinks and unknown out-of-the-way places. Some resting on the bare earth may perchance be seized by some blind worm, and made to close the door of its lowly habitation, and, germinating there, may, in after years, when all who now live have passed away, spread its huge arms, and afford a grateful shelter to those who are to come after us. Just so the broad sheet you daily publish conveys to every civilised part of the world the thoughts and sentiments of those who lead and form public opinion, while it never fails to give the latest expression of science, literature, and art. Much of all this may, like the flying plane tree seeds, fall on unproductive soil; yet who shall say in that ceaseless stream of intelligence how many a sympathetic chord of the human heart may be touched, or how many thoughts and sentiments so imbibed may germinate, and, gaining strength with years, may change the whole current of a life, and form the statesman, the scientist, or the man of letters? Thus musing it occurred to me that the statistical results I had arrived at might, perhaps, interest some other boys than those for whom they were intended, and if thought worthy of a place in *The Times* might inspire a more than passing interest in an otherwise most uninviting subject.

"The statistics of the coal trade show that during the year 1881 the quantity of coal raised in Great Britain was no less than 154,184,300 tons. When the eye passes over these nine figures it does not leave on the mind a very vivid picture of the reality— it does not say much for the twelve months of incessant toil of the 495,000 men who are employed in this vast industry; hence I have endeavoured in a pictorial form to convey to the mind's eye of my young friends something like the true meaning of those figures; for mere magnitude to the youthful mind has always an absorbing interest, and the gigantic works of the ancients fortunately supply us with a ready means of comparison with our own. Let us take as an example the great pyramid of Geeza, a work of human labour which has excited the admiration of the world for thousands of years. Though in itself inaccessible to my young friends, we fortunately have its base clearly marked out in the metropolis.

"When Inigo Jones laid out the plan of Lincoln's-inn-fields he placed the houses on opposite sides of the square just as far from each other as to inclose a space between them of precisely the same dimensions as the base of the great pyramid. Measuring up to the front walls of the houses this space is just equal to 11 acres and 4 poles. Now, if my young friends will imagine St. Paul's Cathedral to be placed in the centre of this square space, and having a flagstaff of 95 ft. in height standing up above the top of the cross, we shall have attained an altitude of 499 ft. which is precisely equal to that of the great pyramid. Further let us imagine that four ropes are made to extend from the top of this flagstaff, each one terminating at one of the four corners of the square and touching the front walls of the houses. We shall then have a perfect outline of the pyramid of exactly the same size as the original. The whole space inclosed within these diagonal ropes is equal to 79,881,417 cubic feet, and if occupied by one solid mass of coal it would weigh 2,781,581 tons—a mass less than $\frac{1}{55}$th part of the coal raised last year in Great Britain. In fact, the coal trade could supply such a mass as this every week, and at the end of the year have more than nine millions of tons to spare.

"Higher up the Nile, Thebes presents us with another example of what may be accomplished by human labour. The great temple of Rameses at Carnac, with its hundred columns of 12 ft. in diameter, and over 100 ft. in height, cannot fail to deeply impress the imagination of all who, in their mind's eye, can realise this magnificent colonnade. It may be interesting to ascertain what size of column and what extent of colonnade we could construct with the coal we laboriously sculpture from its solid bed in every year.

"Let us imagine a plain cylindrical column of 50 ft. in diameter, and of 500 ft. in height, our one year's production of coal would suffice to make no less than 4,511 of these gigantic columns, which, if placed only at their own diameter apart, would form a colonnade which would extend in a straight line to a distance of no less than 85 miles and 750 yards—in fact, we dig in every working day throughout the year a little more than enough to form fourteen of these tall and massive columns, which, if placed upon each other, would reach an altitude of 7,000 ft.

"But there is yet another great work of antiquity which our boys will not fail to remember as offering itself for comparison; they have all heard of the Great Wall of China, which was erected more than 2,000 years ago to exclude the Tartars from the Chinese Empire. This great wall extends to a distance of 1,400 miles, and is 20 ft. in height, and 24 ft. in thickness, and hence contains no less than 3,548,160,000 cubic feet of solid matter. Now, our last year's production of coal was 4,427,586,820 cubic feet, and is sufficient in bulk to build a wall round London of 200 miles in length, 100 ft. high, and 41 ft. 11 in. in thickness—a mass not only equal to the whole cubic contents of the Great Wall of China, but sufficient to add another 346 miles to its length.

"These imaginary coal structures can scarcely fail to impress the mind of youth with the enormous consumption of coal; and when they are told that in many of its applications the useful effect obtained is not one-fifth of its theoretic capabilities, they will be enabled to form some idea of the vast importance of the economic problem which calls so loudly for solution. They must not, however, fall into the too common error of supposing that the electric light by superseding gas is to do away with the use of coal in the production of light, or that dynamo-electric machines will largely replace the steam-engine and boiler.

"Although coal is still our great agent in the production of motive power, it must not be forgotten that Sir William Thomson has clearly shown that by the use of dynamo-electric machines, worked by the falls of Niagara, motive power could be generated to an almost unlimited extent, and that no less than 26,250 horse-power so obtained could be conveyed to a distance of 300 miles by means of a single copper-wire of half-an-inch in diameter, with a loss in transmission of not more than 20 per cent., and hence delivering at the opposite end of the wire 21,000 horse-power.

"What a magnificent vista of legitimate mercantile enterprise this simple fact opens up for our own country. Why should we not at once connect London with one of our nearest coalfields by means of a copper rod of one inch in diameter and capable of transmitting 84,000 horse-power to London, and thus practically bring up the coal by wire instead of by rail.

"Let us now see what is the equivalent in coal of this amount of motive power. Assuming that each horse-power can be generated by the consumption of 3 lbs. of coal per hour, and that the engines work six and a half days per week, we should require an annual consumption of coal equal to 1,012,600 tons to produce such a result.

"Now all this coal would in the case assumed be burned at the pit's mouth at a cost of 6s. per ton for large and 2s. per ton for small coal—that is, at less than one-fourth the cost of coal in London. This would immensely reduce the cost of the electric light, and of the motive power now used in London for such a vast variety of purposes, and at the same time save us from the enormous volumes of smoke and foul gas which this million of tons of coal would make if burned in our midst. A one-inch diameter copper rod would cost about 533*l.* per mile, and if laid to a colliery 120 miles away, the interest at 5 per cent. on its first cost would be less than 1*d.* per ton on the coal practically conveyed by it direct into the house of the consumer."

It would require nearly as fine an exercise of imagination as is displayed in the above extracts to adequately represent the service which the Bessemer process has rendered to the industrial world. According to the best information extant it appears that in the twenty-one years that elapsed after the process was first successfully worked, the production of steel by it, notwithstanding its slow progress at first, amounted to no less than 25,000,000 tons ; and if we were to estimate the saving, as compared with the old process which it superseded, at 40*l.* a ton, the total would be about 1,000,000,000*l.* In 1882 the world's production was over 4,000,000 tons. Over 100 works had adopted it, and over 3,300 converters had been erected.

Such a man needs no honours; but no industrial nation could afford to let him go unhonoured. Hence honours and distinctions have been showered upon him from all quarters. In recognition of the value of his process he was presented with the freedom of the city of Hamburg. The King of Wurtemburg presented him with a gold medal accompanied by a complimentary letter of acknowledgment. The Emperor of Austria, who took a great interest in the progress of the Bessemer process, conferred on him the honour of Knight Commander of the Order of Francis Joseph, accompanying the jewelled cross and circular collar of the Order by a complimentary letter. In 1867 a scientific commission in Paris, in reporting to the Emperor Napoleon III. upon the progress and importance of the Bessemer process, suggested that his Majesty should confer upon its inventor the Grand Cross of the Legion of Honour. The Emperor assented on condition that the English minister in Paris would permit Sir Henry to wear it. The English Government refused this permission ; but on the occasion of the next Paris Exhibition the Emperor presented him personally with a magnificent gold medal, weighing twelve ounces, in recognition of the value of his inventions—a recognition which appears all the more spontaneous because Sir Henry was not an exhibitor. The Society of Arts and Manufactures at Berlin also presented him with a gold medal, and made him an honorary member. Nor were these all his foreign honours. In the United States, where the Bessemer process is about as extensively worked as in the United Kingdom, titles, ribbons, and decorations are unknown, but in the State of Indiana, in a district rich in anthracite coal and pure iron ore, a new scene of industry is springing into existence. Furnaces are in full blast, houses and factories are being built, and the nucleus of a great town already exists, to be known in future by the name of Bessemer.

While he was without honour at home he was not without emolument. When his patent expired in 1870 he found that he had received in royalties over a million sterling, or, to use his own expression, 1,057,748 "of the beautiful little gold medals which are issued by the Royal Mint, with the benign features of her most gracious Majesty Queen Victoria stamped upon them." His Sheffield works were a source of unexampled profit. When the partnership expired it was found that the firm had divided in profits, during their fourteen years' working, fifty-seven times the amount of the subscribed capital, and after that the works, which had been considerably extended at the expense of revenue, were sold for twenty-four times the amount of the whole sub-scribed capital. They thus received altogether eighty-one times their original capital in fourteen years. In other words, their profits for every two months amounted to as much as the capital originally invested in the business.

In later years he received honours even in his own country. In 1872 the Albert gold medal was presented to him by the Prince of Wales for his eminent services to arts, manufactures, and commerce in developing the manufacture of steel. The presentation was made personally at Marlborough House; and it is remarkable that this was the first national recognition of his services. The previous recipients of the Albert medal were Sir Rowland Hill, the Emperor Napoleon III., Professor Faraday, Sir Charles Wheatstone, Sir Joseph Whitworth, Baron Liebig,

M. de Lesseps, and Sir Henry Cole. In the following year (1873) Sir Henry established a gold medal under the auspices of the Iron and Steel Institute, to be awarded annually by that body to persons distinguished by their inventions or services in promoting the manufacture of iron or steel. In 1878 the Institution of Civil Engineers conferred on him the honorary title of C.E., and awarded him the Howard Quinquennial Prize, the highest they could bestow. In 1879 he was elected a Fellow of the Royal Society. In 1879, too, the honour of knighthood was conferred on him by the Queen. *A propos* of this honour, the President of the Iron and Steel Institute remarked that the name of Bessemer was one that ought not to be exchanged for the greatest title ever given to man.

In 1880 the freedom of the city of London was presented to him. The gold casket in which it was inclosed illustrated the process of steel-making from the conversion of the raw material to the application of the steel. It was of solid English design, surmounted by a finely modelled figure of Commerce standing between a stack of pig iron and the converter, and thus commending the invention on account of the impetus given by cheap steel to commercial enterprise. Overflowing cornucopiæ at the base signified this success. On either side of the rounded cover were vignettes (in *repoussé* work) of a London and North-Western Railway locomotive (entirely constructed of steel and standing on steel rails) and of a steel clad ship. The two curved ends contained the enamelled arms of the city, with the dragons modelled in high relief. On the centre panel was the medal given annually by Sir Henry Bessemer to the Iron and Steel Institute. The inscription was on the reverse, shields for the Bessemer arms and monograms completing the whole, which rested on a plateau of Bessemer steel.

In making the presentation, the City Chamberlain said to Sir Henry: "I find that with the exceptions of Dr. Jenner, who introduced the practice of vaccination, and of the late Sir Rowland Hill, the originator of the uniform prepaid penny postal system, you are the only great discoverer who has received this honorary freedom. The annals of human progress in the arts furnish few parallels to the revolution which has been effected by the invention with which your name will be ever associated. It has secured that the 'iron age' shall not return again, for that metal has already succumbed to its competitor and inevitable successor. Our guns, large and small, in order to perform the work of penetration required of them, must be constructed henceforth of steel; our ships of war must be steel-ribbed and steel-clad; and the contest between iron and iron, waged hitherto in the attempt to solve that which appears to be insoluble, will have to be waged in future between steel shot and steel armour plates. But it is in the arts of commerce and of peaceful life that the revolution which you have effected will be more and more felt. The world now runs upon wheels, and with greater speed, safety, and more comfort than heretofore, by reason of steel-tyred wheels without a weld, and polished steel rails, produced at one-sixth of the cost of that metal prior to your discovery. Our locomotives are now built of homogeneous steel, while viaducts, bridges, merchant vessels, anchors, boilers, and other parts of moto-machinery, as well as the thousand appliances and conveniences of civilised life, are now to a great extent constructed of the modified iron which you first introduced."

SIR WILLIAM SIEMENS.

CHAPTER VI.

" What cannot art and industry perform
When science plans the progress of their toil ! "

BEATTIE.

SIR CHARLES WILLIAM SIEMENS, D.C.L., LL.D., F.R.S., is a member of a family eminent for their scientific knowledge and practical skill. The possession of such unusual talents by a whole family is a rare occurrence in the intellectual life of England ; but it has not been so exceptional in Germany. At present, however, the Brothers Siemens are the most conspicuous examples of this sort of constellation of genius ; and, although natives of Germany, they have so distributed their talents that England, Russia, and the Fatherland have been made the scene of their labours, and have been able to claim the distinction of being the birthplace of their respective discoveries. It is a common impression in England that if Nature bestows her gifts in more than ordinary profusion upon one individual she withholds them from others of the same family. Germany is the country which has been most fruitful in examples to the contrary ; and in applied science her latest contribution of a family of scientists has imparted lustre and celerity to the scientific and industrial progress of three of the greatest nations in the world. The steel trade claims one of them among her greatest inventors.

In one respect the labours of this inventor present an instructive
contrast to those of most English inventors. " All is race " was
the favourite saying of an English statesman in reference to the
peculiarities of national character, and it is not difficult to discern
some prominent characteristics of their race in the scientific career
of the Brothers Siemens. Guizot says that, "When under some broad
point of view, or under some essential relation, any principle appears
to the Germans good and beautiful, they conceive for it an exclusive
admiration and sympathy. They are generally inclined to admire
and be overcome with passion ; the imperfections, the interruptions
of a bad state of things strike them but little. Singular contrast !
In the purely intellectual sphere, in the research for and combination
of ideas, no nation has more extension of mind, more philosophic
impartiality." " The French," says Hegel, "call the Germans
entiers—' entire '—*i.e.* stubborn ; they are also strangers to the
whimsical originality of the English. The Englishman attaches
his idea of liberty to the special [as opposed to the general] ; he
does not trouble himself about theoretic conclusions ; but, on the
contrary, feels himself so much the more at liberty the more his
course of action or his license to act contravenes, or runs counter
to, theoretic conclusions or general principles." The sympathy of
the German mind for general principles, and the tenacity with
which it clings to them, are amply illustrated in the life of Sir
William Siemens. His inventions are the embodiment or practical
application of scientific principles ; and the devotion with which
he worked for years at the self-imposed task of utilising one after
another the sort of ideas which Englishmen are apt to stigmatise
as theories, is perhaps unsurpassed in the history of English
inventions. There are instances in which an Englishman may
have laboured as persistently, and even more exclusively, for the
attainment of one definite end by any or all sorts of means ; but
it would be difficult to find in English history an instance in which
an inventor has been so confident of the possible utility of a few

general principles that he has worked out from them several great inventions.

In order to compare the different standpoints from which two great electricians trace their inventive inspirations, it is only necessary to read the two following sentences. The secret of Mr. Edison's success in the invention of electrical apparatus, according to an English literary journal,[1] may be thus summed up in his own words: "Whenever by theory, analogy, and calculation I have satisfied myself that the result I desire is impossible, I am then sure that I am on the verge of a discovery." On the other hand, Sir William Siemens says : "The further we advance, the more thoroughly we approach the indications of pure science in our practical results." No man in England has done more than himself towards this end.

Hence he is more than a mere inventor. Professor Forbes, writing at the very time when Sir William Siemens was enter·ing upon the series of experiments that initiated his greatest works, said : "Watt's parallel motion, perhaps the most ingenious of his inventions, would not have made a great reputation; nor does the endless variety of machines used in the arts, as in spinning, printing, and paper-making, stand higher. It is when the inventor places matter in new relations to force, or derives power from new sources, or teaches heat and electricity to act under new conditions, that he becomes really a Mechanical Philosopher." Sir William Siemens is indeed a Mechanical Philosopher.

Born at Lenthe in Hanover, on April 4, 1823, his early education was acquired at Lübeck, where the German guild system appears to have attracted his attention, for he repeatedly referred to it in after life. "When a boy at school," he says, " I was living under the full vigour of the old Guild system. In going through the streets of Lübeck I saw Carpenters' Arms, Tailors' Arms, Goldsmiths'

[1] The *Athenæum.*

Arms, and Blacksmiths' Arms. These were lodging houses where every journeyman belonging to that trade or craft had to stop if he came into the town. In commencing his career, he had to be bound as an apprentice for three or four years; and the master, on taking an apprentice, had to enter into an engagement to teach him the art and mystery, which meant the science of his trade. Before the young man could leave his state of apprenticeship he had to pass a certain examination; he had to produce his *Gesellen-stück* or journeyman piece of work, and if that was found satisfactory, he was pronounced a journeyman. He had then to travel for four years from place to place, not being allowed to remain for longer than four months under any one master; he had to go from city to city, and thus pick up knowledge in the best way that could have been devised in those days. Then, after he had completed his time of travel, on coming back to his native city, he could not settle as a master in his trade until he had produced his *Meister-stück*, or master-piece. These master-pieces in the trade were frequently works of art in every sense of the word. They were, in blacksmithy, for instance, the most splendid pieces of armoury; in every trade, and in clocks above all others, great skill was displayed in their production. These were examined by the Guild Masters' Committee, and upon approval were exposed at the Arms of the Trade for a certain time, after which the journeyman was pronounced a master; he was then allowed to marry, provided he had made choice of a young woman of unimpeachable character. These rules would hardly suit the taste of the present day, but still there was a great deal of good in those old Guild practices." This system was abolished in Germany in 1869, but the stimulus it afforded to excellence of workmanship appeared to have made an early and lasting impression on his mind.

Leaving Lubeck, he proceeded to the Polytechnical school at Magdeburg, with the view of acquiring a knowledge of physical science. It is natural to ask what were the facilities afforded in early life to one who has since become one of the most liberal advocates of technical education, for acquiring his own knowledge of science and mechanics ; the more so as he maintains that every school ought to possess a laboratory, not necessarily involving a large expenditure for apparatus, because the most instructive apparatus is that which is built up in the simplest possible manner by means of pulleys, cords, wires, and glass tubes, and which, therefore, calls into requisition the constructive ingenuity of the student himself. In point of fact, his own facilities for acquiring such knowledge were so scanty, compared with those generally provided in laboratories now, that, as he has himself stated, on carrying his thoughts back to the physical laboratory connected with the school where he received his scientific instruction it would almost seem impossible that anything efficient could have been taught there. For example, the appliances then at his command for acquiring the rudiments of that science—electricity—in which he was afterwards to become so great a master, were of a very primitive kind. They consisted of a battery composed of flannel and some pieces of copper piled up to a certain height, so as to produce a spark ; then there was a long scale with a pulley to show the acceleration of a body by gravitation, and also an electrical friction-machine such as may now be seen in an advanced nursery. These were the only scientific apparatus then to be seen at the school which produced one of the first and greatest inventors of electrical appliances.

From that school he went to Göttingen University ; and while studying there an event occurred which had an important bearing on his subsequent career, and which he has himself described as the determining incident of his life. Here is his own account of it : " At that time (1841) that form of energy known as the electric current was nothing more than the philosopher's delight. Its first practical application might be traced to the town of

Birmingham, where Mr. George Elkington, utilising the discoveries of Davy, Faraday, and Jacobi, established a practical process of electro plating in 1842. It affords me great satisfaction to be able to state that I had something to do with that first practical application of electricity, for in March of the following year (1843) I presented myself before Mr. Elkington with an improvement of his process, which he adopted, and in so doing gave me my first start in practical life. When the electro-type process first became known it excited a very general interest; and although I was only a young student at Göttingen, under twenty years of age, who had just entered upon his practical career as a mechanical engineer, I joined my brother, Werner Siemens, then a young lieutenant of artillery in the Prussian service, in his endeavours to accomplish electro-gilding, the first impulse in this direction having been given by Professor C. Himley, then of Göttingen. After attaining some promising results, a spirit of enterprise came over me so strong that I tore myself away from the narrow circumstances surrounding me, and landed at the East-end of London with only a few pounds in my pocket, and without friends, but with an ardent confidence of ultimate success within my breast. I expected to find some office in which inventions were examined, and rewarded if found meritorious; but no one could direct me to such a place. In walking along Finsbury Pavement, I saw written up in large letters so and so (I forget the name) 'undertaker' and the thought struck me that this must be the place I was in quest of. At any rate, I thought that a person advertising himself as an undertaker would not refuse to look into my invention, with a view of obtaining for me the sought-for recognition or reward. On entering the place I soon convinced myself, however, that I had come decidedly too soon for the kind of enterprise there contemplated, and finding myself confronted with the proprietor of the establishment, I covered my retreat by what he must have thought a very inadequate excuse. By dint of perseverance I found my way to the patent office of Messrs. Poole and Carpmael, who received me kindly, and provided me with a letter of introduction to Mr. Elkington. Armed with this letter, I proceeded to Birmingham to plead my cause with him. In thinking back to that time, I wonder at the patience with which Mr. Elkington listened to what I had to say, being very young, and scarcely able to find English words to convey my meaning. After showing me what he was doing already in the way of electro-plating, Mr. Elkington sent me back to London in order to read some patents of his own, asking me to return if, after perusal, I still thought I could teach him anything. To my great disappointment I found that the chemical solutions I had been using were actually mentioned in one of his patents, although in a manner that would hardly have sufficed to enable a third person to obtain practical results. On my return to Birmingham I frankly stated what I had found, and with this frankness I evidently gained the favour of Mr. Josiah Mason, who had just joined Mr. Elkington in business, and whose name as Sir Josiah Mason will ever be remembered for his munificent endowment of education. It was agreed that I should not be judged by the novelty of my invention, but by the results which I promised—namely, of being able to deposit with a smooth surface 3dwt. of silver upon a dish-cover, the crystalline structure of the deposit having theretofore been a source of difficulty. In this I succeeded, and I was able to return to my native country and my mechanical engineering a comparative Crœsus. Notwithstanding the leaps of time," he said nearly forty years afterwards, " my heart still beats quick each time I come back to the scene of this, the determining incident of my life."

In 1843 he became a pupil in the engine works of Count Stolberg, where he intended to acquire the workshop knowledge necessary to a mechanical engineer. While thus employed he worked out another invention. This was a steam engine governor, an invention which was to some extent suggested by his elder brother, but which was perfected by himself. It was a decided advance upon Watt's centrifugal governor. In Watt's the rotation of the pendulum varied with every change in the relative condition of the power and load of the engine; and in consequence of this dependence and defects of construction it was, correctly speaking, only a moderator. To avoid these imperfections the Brothers Siemens made their governor with the pendulum independent in its action of change in its rotation; and it was provided always with a store of power ready to overcome the resistance of the valve at the first moment when the balance between the power and load of the engine was disturbed. Between the pendulum and a wheel driven by the engine there was a differential motion which acted instantly upon the supply valve whenever a sudden disturbance of balance took place, and it possessed the power of maintaining the regularity of the machinery at the same speed when the load reached its maximum as well as when it was at its minimum.

With this invention he returned to England in 1844, and soon determined to stay here. His object in doing so was to enjoy the security which the English patent law affords to inventors. In his own country there were then no such laws: privileges were sometimes granted by the Government to applicants for a very short period; but even this limited protection had been so often refused to great inventions and granted to inventors of small mechanical improvements, that he determined to prosecute his labours in England, where the law granted at least fourteen years' exclusive protection to all sorts of inventions. The chronometric governor, though less successful, commercially speaking, than the first invention, was the means of bringing him into contact with the engineering world. In course of time it was applied by Sir George Airy, then the Astronomer Royal, for regulating the motion of the great transit and touch-recording instruments at the Royal Observatory, where it still continues to be employed.

Another early invention of the two brothers was the art of "Anastatic printing," which in 1845 was made the subject of a lecture by Professor Faraday before the Royal Institution. By this process old or new printed matter could be reproduced. The method of treatment consisted in first applying caustic baryta or strontia to the printed matter, in order to convert the resinous ingredients of the printing ink into a non-soluble soap, and next applying sulphurous acid to precipitate the stearine. By this means the printed matter, on being subjected to pressure, could be transferred to zinc. Both these inventions, although very ingenious, did not come largely into use, and hence were not very profitable to the inventors.

In 1846 Sir William Siemens constructed an air-pump which was well received, and in 1851 a water meter which has since been in general use. The latter, with subsequent modifications, was found to suit all circumstances of varying pressure and to give early intimation of leakage. It has been largely manufactured and used both in England and on the Continent.

Meanwhile he had entered upon a field of study more extensive in its range and more fruitful in its results. In 1846 he began to study the economy of fuel in the light of recent investigations respecting the true nature of heat. Three or four years previously some scientists, working far apart, had independently developed the theory that heat is a manifestation of motion between the different particles of matter, and that it can be expressed in equivalent values of palpable motion. This was called the mechanical equivalent of heat. The first complete experimental demonstration of this immateriality of heat was made by Davy, who melted two pieces of ice by rubbing them together in an atmosphere below the freezing point. Several other experiments having a similar effect led Mayer in 1842 and

Joule in 1843 to assert that heat is the equivalent for work spent in agitating a fluid; and the subsequent experiments of Joule were then considered by many eminent scientific men to have established this principle beyond doubt. In 1849 Joule arrived at the determination, since then universally adopted, of the numerical relation between the units of heat and units of work, and thus expressed the mechanical value of heat. He established the rule, now in general use, that 772 foot pounds of work (that is, 772 times the amount of work required to raise a weight of 1lb. through a space of 1ft.) is required to generate as much heat as will raise the temperature of a pound of water by one degree.

This subject was one that early engaged the attention of the Brothers Siemens; and at the age of twenty-three Sir William adopted the new theory. He read the treatises of Joule, Carnot, and Mayer, and having thus mastered the available knowledge of the greatest authorities on the subject, he proceeded to experiment on the principles then brought to light. On comparing the theoretic power of heat with the mechanical power given off by the heat applied to steam engines and caloric engines generally, he saw that there was a large margin for improvement. He at once determined to try to save or utilise some of this wasted heat; and conceived the idea of making a regenerator or accumulator for the purpose of retaining a limited quantity of heat and capable of yielding it up again when required for the performance of any work. Accordingly in 1847 he constructed in the factory of Mr. John Hick, of Bolton, an engine of four horse-power which had a condenser provided with regenerators, and which attained partial success by the use of superheated steam. The economy of fuel was considerable; but this saving was attended with mechanical difficulties which at that time he was unable to solve. He did not, however, abandon the subject. Continuing his examination of wasted heat in steam engines, he wrote a paper for the Institution of Civil Engineers in 1852 " On the Conversion of Heat into Mechanical Effect." At that time very little effort had been made to apply the new principle. Both Stirling in Scotland and Ericsson in America invented air engines for the application of the mechanical theory; but in practical working both inventions failed. In his paper before the Civil Engineers Sir William endeavoured to set forth the probable causes of the failure of Ericsson's experiment in America, and to guard against a sweeping condemnation on that account of some of the means that Ericsson had employed. He also suggested that by carrying the principle of the expansion of heat to a much further extent than had been done up to that time a very considerable proportion of the theoretical duty could be realised from coal, and that future progress would probably be in the direction of this extended application of expansive working. To overcome the mechanical difficulties that attended his own first attempts to utilise some of this wasted heat, he, along with his younger brother Frederick, turned his attention to furnace-heat, and entered upon a costly and prolonged series of experiments with the view of utilising in a practical way this theoretic power. The mechanical theory continued to advance in principle but not in practical application. To illustrate the theory Sir William Thompson, applying it to the mechanical actions of living creatures, said it appeared certain, from the most careful physiological researches, that a living animal had not the power of originating mechanical energy; and that all the work done by a living animal in the course of its life, and all the heat that had been emitted from it, together with the heat that would be obtained by burning the combustible matter which had been lost from its body during its life, and by burning its body after death, made up together an exact equivalent to the heat that would be obtained by burning as much food as it had used during its life, and an amount of fuel that would generate as much heat as its body if burnt immediately after death.

To prove "the imperishable nature of physical forces and their mutual convertibility," Sir William Siemens used the following simple illustration. He said that a weight falling over a pulley, to which it was attached by a string, would impart rotatory motion to a fly-wheel fixed upon the same axis with the pulley, and the velocity imparted to the wheel would cause the string to wind itself upon the pulley till the weight had reached nearly its original elevation. If the friction of the spindle and the resistance of the atmosphere could be dispensed with, the weight would be lifted to precisely the same point from whence it fell before the motion of the wheel was arrested. In descending again, it would impart motion to the wheel as before, and this operation of the weight, of alternately falling and rising, could continue *ad infinitum.* If the string were cut at the instant when the weight had descended, the rotation of the wheel would continue uniformly, but it might soon be brought to a stop by immersing it in a basin filled with water. In this case the water was the recipient of the force due to the falling weight residing in the wheel, and by repeating the same experiment a sufficient number of times we should find an increase of temperature in the water. If the weight falling over a pulley were one pound, and the distance through which it fell one foot, then each impulse given to the wheel would represent one foot pound, our commonly adopted unit of force; and if the water contained in the basin weighed also one pound, it would require 770 repetitions of the experiment of arresting the wheel in the water before the temperature of that water was increased by one degree Fahrenheit.

Sir William Siemens' experiments in the practical application of this new theory resulted in the construction of a regenerative steam engine, which effected such a saving of fuel that several of them were soon put into practical operation in England, France, and Germany. These, varying from five to forty horse-power, were regarded as having proved the practicability of the principle involved, although their inventor admitted they were still capable of improvement.   In 1856 he explained his engine to the Royal Institution.   He stated that it was the result of ten years' experimental researches, and it was, he thought, the first practical application of the mechanical theory of heat, of which he was proud to call himself an early disciple.   Others, he said, more able than himself, might probably have arrived sooner at a practically useful result, but he claimed for himself at least that strong conviction, approaching enthusiasm, which alone could have given him strength to combat successfully the general discouragement and the serious disappointments he had met with.   One of his steam engines of twenty horse-power was placed in the Paris Exhibition of 1857; it did not fully answer the inventor's expectations; but another of seven horse-power being substituted, it was found to work with good economical results.

Viewed in the light of the new theory, he explained that the heat given out in the condenser of a steam engine represented a loss of mechanical effect amounting to $\frac{13}{14}$ of the total heat imparted to the boiler, the remaining $\frac{1}{14}$ part being all the heat really converted into mechanical effect.   The greater portion of the heat lost might be utilised by a perfect engine.   A vast field for practical discovery was thus opened out ; but if it were asked whether it was worth while to leave the tried and approved forms of engines then in use to seek for economy, however great, in a new direction, considering the vast extent of our coalfields, he replied that the coal in its transit from the pit to the furnace acquired a considerable value, which in this country might be estimated at 8l. per horse-power per annum (taking a consumption of $13\frac{1}{2}$ tons of coal at an average expenditure of 12s. a ton).   Estimating the total force of the stationary and locomotive engines then employed in this country at one million nominal horse-power, it followed that the total expenditure for steam coal amounted to eight million pounds sterling per annum, of which at least two-thirds might be saved. In other countries, where coal was scarce, the importance of economy was still more apparent; but it was of the highest importance for marine engines, the coals for which had to be purchased at transatlantic stations at a cost of several pounds per ton, not to mention the indirect cost of its carriage by the steamer itself in place of merchandise. Such were the economic considerations which led him to persevere with his experiments.

The engine he made for this purpose had three cylinders so constructed that the steam took up heat and gave it out as it passed from one to the others. Two, called working cylinders, had plungers, and the other had a piston. The steam was heated to a high temperature in the working cylinders, under each of which there was a fire; and after being partly consumed in doing the mechanical work of lifting the plungers, it passed into the regenerator or respirator. The regenerator was an invention of Dr. Stirling's, who discovered that if heat be passed through a compartment filled with sieves of wire gauze, or even minutely divided passages, it will leave a large amount behind. When, therefore, the steam in the new engine reached the regenerator, it had to traverse a mass of metallic wire gauze or plates, called the respirator, where its temperature was thus raised from 250° to 600° or 700° Fahr. In consequence of the addition of temperature that the steam received in its passage through the respirator, its elastic force was doubled. It then returned to the plunger cylinders, where it received additional temperature and commenced its round again. Thus the same steam was continually employed in going round and round again, depositing on its regress through the respirator the heat it had received on its egress through the same, less only the quantity which had been lost in its expansion below the working piston, and which was converted into mechanical effect

L

or engine power. The expansion and simultaneous reduction of the temperature of the steam caused a diminution of its pressure from four to nearly one atmosphere ; so that, while one working plunger could effect its return stroke without opposing pressure, the second plunger made its effective or outward stroke impelled by a pressure of four atmospheres. These plungers were connected with the working crank of the engine in the usual way. The quantity of fresh steam admitted from the boilers into the regenerative cylinder at each stroke did not exceed one-tenth of the steam contained in the working cylinders of the engine.

The respirator, according to Sir William Siemens, fulfilled its office with surprising rapidity and perfection, if it were made of suitable proportions. It had been applied without success to hot-air engines by Stirling and Ericsson, but failed for want of proper application ; for it had been assumed (in accordance with the mechanical theory of heat), that it was capable of recovering all the heat imparted to the air ; and in consequence no sufficient provision of heating apparatus had been made. It having been found impossible to produce what in effect would have been a perpetual motion, the respirator had then been discarded entirely, and even in 1856 it was looked upon with great suspicion by engineers and men of science. Sir William Siemens was still confident, however, that its real merits to recover heat that could not practically be converted by one single operation into mechanical effect would be better appreciated. The rapidity with which the temperature of a volume of steam was raised from 250° to 650° Fahr. by means of a respirator was shown by the fact that he had obtained with his engines a velocity of 150 revolutions per minute. The single action of heating the steam occupied only a quarter of the time of the entire revolution of the engine and it followed that it was accomplished in one-tenth part of a second. In explanation of this phenomenon, it was contended that the transmission of a given amount of heat from a hotter to a cooler body was proportionate to the heating surface multiplied by the time occupied, and that the latter factor might be reduced *ad libitum* by increasing the former proportionately. The air engines of Stirling and Ericsson had also failed because their heated cylinders had been rapidly destroyed by the fire ; but the cause of this was that an insufficient extent of heating surface had been provided. Sir William Siemens' experience led him to believe that his heating vessels would last from three to five years ; and being only a piece of rough casting that could be replaced in a few hours, at a cost below that of a slight boiler repair, he thought he had practically solved the difficulty arising from high temperature.

But practical working dispelled these hopes. His engine is still admired as an ingenious application of the mechanical theory of heat, and is admitted to be capable of economising fuel ; but the wear and tear of the heating vessels were found to be too great for ordinary working. Hence the invention soon fell into disuse, but the faith of the inventor in its principle remained steadfast.

In his presidential address to the Institute of Naval Architects in 1882, the Earl of Ravensworth stated that during the nine years up to 1872 the improvements in marine engineering were so vast that the consumption of fuel was reduced by one-half. Since that time by the use of combined engines a saving of 13 per cent. was effected, and subsequently by the introduction of the triple engine a further saving had been made of 12 per cent., making altogether a total saving of 75 per cent. Notwithstanding these facts, Sir William Siemens told the British Association in August, 1882, that "the best steam-engine yet constructed does not yield in mechanical effect more than one-seventh part of the heat energy residing in the fuel consumed. To obtain more advantageous primary conditions we have to turn to the caloric or gas engine, in which we have also to make reductions from the theoretical efficiency, on account of the rather serious loss of heat by absorption into the working cylinder, which has to be cooled artificially in order to keep its temperature down to a point at which lubrication is possible; this, together with frictional loss, cannot be taken at less than one-half, and reduces the factor of efficiency of the engine to one-fourth. But the gas or caloric engine combines the conditions most favourable to the attainment of maximum results, and it may reasonably be supposed that the difficulties still in the way of their application on a large scale will gradually be removed. Before many years have elapsed we may find in our factories and on board our ships engines with a fuel consumption not exceeding one pound of coal per effective horse power per hour, in which the gas producer takes the place of the somewhat complex and dangerous steam boiler. The advent of such an engine and of the dynamo machine must mark a new era of material progress at least equal to that produced by the introduction of steam power in the early part of our century."

While the regenerative engine was for the time apparently abandoned, the inventor's efforts to apply the mechanical theory of heat to industrial purposes were not. In 1857 his brother Frederick suggested to him the employment of regenerators for the purpose of getting up a high degree of heat in furnaces, and he thenceforth laboured to attain this result. In the course of the next four or five years he constructed several different forms of furnaces, each successive design embodying the improvements that experience suggested. His first furnaces were used for heating bars of steel, and they performed this operation with encouraging success. But in attempting to apply the principle to larger furnaces serious practical difficulties arose, which for a time appeared insurmountable. Eventually he tried the plan of volatilising the solid fuel, and in this way he succeeded. By first converting coal into gas and then using it in the gaseous form in regenerators, he obtained practical results surpassing even his own most sanguine expectations. His first furnaces, which were erected at Sheffield and Manchester, were imperfect both in principle and construction, and to overcome their defects the design was elaborated till the mechanism became too intricate to be intrusted to ordinary workmen. After the conversion of the fuel into gas, he discovered that another essential improvement in the construction of the new furnace was the complete separation of the fire-place or gas producer from the heating chambers or furnace. A regenerative furnace with this improvement was constructed at a glass works near Birmingham in 1861, and it was found to be simple in its operation and economical in its results. Several others were erected shortly afterwards on the same principles, the designs in each case being furnished by Sir William Siemens.

In the year 1862 events occurred that had the effect of prominently directing public attention to the success which then crowned his labours in the practical application of the mechanical theory of heat. In June of that year two of the most popular authorities in England gave lectures at the Royal Institution on the results of the labours of the two German scientists who had done most for the development of that theory ; and the careers of these two Germans, as narrated by these two lecturers, presented an instructive, if not a tragic, contrast. In concluding a graphic exposition of the mechanical theory of heat, Professor Tyndall said : " To whom are we indebted for the striking generalisations of this evening's discourse ? All that I have laid before you is the work of a man of whom you have scarcely ever heard. All that I have brought before you has been taken from the labours of a German physician, named Mayer. Without external stimulus, and pursuing his profession as town physician in Heilbronn, this man was the first to raise the conception of the interaction of natural forces to clearness in his own mind. And yet he is scarcely ever heard of in scientific lectures, and even to scientific men his merits are but partially known. Led by his own beautiful researches, and quite independent of Mayer, Mr. Joule published his first paper on the 'Mechanical Value of Heat' in 1843; but in 1842 Mayer had actually calculated the mechanical equivalent of heat from data which a man of rare originality alone could turn to account. From the velocity of sound in air Mayer determined the mechanical equivalent of heat. In 1845 he published his memoir on 'Organic Motion,' and applied the mechanical theory of heat in the most fearless and precise manner to vital processes. He also embraced the other natural agents in his chain of conservation. When we consider the circumstances of Mayer's life and the period at which he wrote, we cannot fail to be struck with astonishment at what he has accomplished. Here was a man of genius working in silence, animated solely by a love of his subject, and arriving at the most important results, some time in advance of those whose lives were entirely devoted to natural philosophy. It was the accident of bleeding a feverish patient at Java in 1840 that led Mayer to speculate on these subjects. He noticed that the venous blood in the tropics was of a much brighter red than in colder latitudes, and his reasoning on this fact led him into the laboratory of natural forces, where he has worked with such signal ability and success. Well, you will desire to know what has become of this man. His mind gave way; he became insane, and he was sent to a lunatic asylum. In a biographical dictionary of his country it is stated that he died there; but this is incorrect. He recovered; and I believe is at this moment a cultivator of vineyards at Heilbronn."

A fortnight later an account was given at the same institution of Sir William Siemens' regenerative gas furnace, the greatest triumph in the practical application of the principles enunciated by Mayer and others. That lecture was delivered by Michael Faraday, the prince of pure experimentalists; and it has the historic interest of being the last lecture he was able to deliver. The circumstances in which it was delivered were memorable.

Some weeks previously Sir William Siemens received the following letter from Faraday: "I have just returned from Birmingham, where I saw at Chance's works the application of your furnaces to glass making. I was very much struck with the whole matter. As our managers want me to end the Friday evenings at the Royal Institution after Easter, I have looked about for a thought, for I have none in myself. I think I should like to speak of the facts I saw at Chance's, if you have no objection. If you assent, can you help me with any drawings, or models, or illustrations, either in the way of thoughts or experiments? Do not say much about it out of doors as yet, for my mind is not settled in what way, if you assent, I shall present the subject."

Sir William Siemens readily assented, and spent two days at Birmingham in showing Faraday over the works where his furnaces were in operation. On the appointed Friday evening, June 20, the venerable *savant* appeared before the Royal Institution for the last time to explain the wonderful simplicity, power, and economy of the regenerative gas furnace. In the course of his lecture, which lasted about an hour, and which he concluded by bidding his audience a pathetic farewell, he accidentally burned his notes; and he was only able afterwards to give the abstract of it that is published in the "Proceedings."

The Siemens regenerative furnace, which was thus brought prominently before the public, consists of three essential parts. The first is the gas producer, which converts the solid fuel into gaseous fuel. A number of these are generally placed outside the works, and the gas produced by them is conducted into the works through underground channels or overhead tubes. Next, there are the regenerators or sunk chambers, which are filled with fire-bricks piled in such a way that a current of air or gas passing through them is broken into a great number of parts, and is checked at every step by the interruption of an additional surface of fire-brick. Four of these chambers are placed below the furnace, and the currents of gas and air can be directed by suitable reversing valves either upwards or downwards through these chambers. Then, thirdly, there is the heated chamber or furnace proper, in which the work of combustion is accomplished. This chamber communicates at each extremity with two of the regenerative chambers : and, on directing currents of gas and air upwards through them, the two gaseous streams meet on entering the heated chamber, where they are ignited. The current then descends through the other two regenerators, and heats them in such a way that while the uppermost chequerwork is heated to nearly the temperature of the furnace, the lower parts are heated to a less and less degree, till at last the products of combustion escape into the chimney comparatively cool. In the course of, say, one hour, the currents are reversed, and the cold air and gas ascending through the two chambers, which have been previously heated, take up the heat there deposited, and again enter into combustion at the entrance into the heated chamber or furnace at nearly the same temperature at which the products of combustion left the furnace, say 500°. By the combustion of these heated gases the heat in the furnace is raised to, say, 1,000°; and after that combustion the remaining products again return through the other two regenerators, heating them as they pass along, and finally escape at the chimney end comparatively cool. By this process of accumulation the most intense temperature can be attained in the furnace chamber without having recourse to gas of high quality or to intensified draught. Practically the limit is reached at the point where the materials of the chamber begin to melt; theoretically the limit exists at the point where combustion ceases, called by Sainte-Claire Deville "the point of dissociation," because at that point (4,500° F.) the two gases—hydrogen and oxygen, which necessarily combine in combustion—become dissociated, showing that combustion only takes place between the limits of about 600° and 4,500° F. It has been found that in a steel melting furnace while the temperature of the melting chamber exceeded 4,000° F., the waste products of combustion escaped into the chimney at 240° F., showing that nearly the whole of the heat generated was absorbed in the furnace in doing its work. This furnace, moreover, has the advantage of preventing smoke, and of using inferior qualities of coal, or such inferior kinds of fuel as peat and lignite.

Faraday especially pointed out the great facility with which these furnaces could be managed. If, he said, while glass is in course of manufacture, an intense heat is required, an abundant supply of gas and air is given ; when the glass is made and the condition has to be reduced to working temperature, the quantity of fuel and air is reduced ; if the combustion in the furnace is required to be gradual from end to end, the inlets of air and gas are placed more or less apart the one from the other. The gas is lighter than the air ; and if a rapid evolution of heat is required, as in a short puddling furnace, the mouth of the gas inlet is placed below that of the air inlet ; if the reverse is required, as in the long tube-welding furnace, the contrary arrangement is used. Not merely can the supply of gas and air to the furnace be governed by valves in the passages, but the very manufacture of the gas fuel itself can be diminished, or even stopped, by cutting off the supply of air to the grate of the gas producer ; and this is important, as there is no gasometer to receive and preserve the aëriform fuel, for it proceeds at once to the furnace. Some of the furnaces have their contents open to the fuel and combustion, as in the puddling and metal-melting arrangements ; others are closed, as in the muffle furnaces and flint-glass furnaces. Because of the great cleanliness of the fuel, some of the glass furnaces, which before had closed pots, now have them open, with great advantage to the working and no detriment to the colour.

Age and experience have not diminished the high estimation in which the regenerative gas furnace is held. After nearly twenty years of continuous working and extended application, Sir Henry Bessemer, in 1880, described it as a beautiful invention which was at once the most philosophic in principle, the most powerful in action, and the most economic of all the contrivances for producing heat by the combustion of coal.

When Sir William Siemens was describing his object in experimenting with his regenerative engine, he said it was impossible to overestimate the benefits that mankind would derive from a motive force at one-third or one-fourth part the cost and incumbrance of the means then in use. But his regenerative gas furnace has been proved capable of making a ton of crucible steel with one-sixth of the fuel required without it. It has been extensively employed in the United Kingdom, on the Continent, and in the industrial centres of the United States. In Russia and Austria, where coal is scarce, peat fuel has been used in the regenerative furnace, and has thus been utilised in the manufacture of steel. In 1868 Sir William Siemens stated that his drawing office could not keep pace with the demand for working drawings of furnaces for iron, steel, zinc, glass, and other works. During that year he was instructed by the Government to reconstruct the furnace department of the Calcutta Mint upon a comprehensive scale; and among other works which underwent a complete transformation in 1868 in accordance with his plans were the Monkbridge Steel and Iron Works, the Hayange and Associated Works, the Llansamlet Zinc Works near Swansea, and the four large plate-glass works of the Marie d'Oignies, Floreffe, Aniche, and Jeumont, involving the construction of 100 puddling furnaces, of steel-melting furnaces (of eighteen and twenty-four pots each), of forge furnaces, and of glass houses of unprecedented dimensions, the practical success of which was complete.

Great, however, as its economic results have been, its inventor looked forward to more remarkable applications of it than any that have ever yet taken place. " It is a favourite project of mine," he says, " which I have not had an opportunity yet of carrying practically into effect,—to place these gas producers at the bottom of coal-pits. A gas shaft would have to be provided to conduct the gas to the surface ; the lifting of coal would be saved, and the gas in its ascent would accumulate such an amount of forward pressure that it might be conducted for a distance of several miles to the works or places of consumption. This plan, so far from being dangerous, would insure a very perfect ventilation of the mine, and would enable us to utilise those waste deposits of small coal (amounting on the average to twenty per cent.) which are now left unutilised within the pit. Another plan of the future which has occupied my attention is the supply of towns with heating gas for domestic and manu-facturing purposes. In the year 1863 a company was formed, with the concurrence of the Corporation of Birmingham, to provide such a supply in that town at the rate of sixpence per 1,000 cubic feet, but the bill necessary for that purpose was thrown out in Committee of the House of Lords, because their lordships thought that, if this was as good a plan as it was represented to be, the existing gas companies would be sure to carry it into effect. It need hardly be said that the existing companies have not carried it into effect, having been constituted for another object." The realisation of this plan was thus indefinitely postponed ; but so far from having lost faith in it, Sir William Siemens told the people of Glasgow in the beginning of 1881 that that town with its adjoining coal-field appeared to be a particularly favourable locality for putting such a plan to a practical trial, and added that when thus supplied with gaseous fuel, the town would not only be able to boast of a clear atmo-sphere, but the streets would be relieved of the most objectionable portion of their daily traffic.

Again, in August, 1882, he told the British Association that he thought "the time is not far distant when both rich and poor will largely resort to gas as the most convenient, the cleanest, and the cheapest of heating agents, and when raw coal will be seen only at the colliery or the gasworks. In all cases where the town to be supplied is within, say, thirty miles of the colliery, the gasworks may with advantage be planted at the mouth, or still better at the bottom, of the pit, whereby all haulage of fuel would be avoided, and the gas, in its ascent from the bottom of the colliery, would acquire an onward pressure sufficient probably to impel it to its destination. The possibility of transporting combustible gas through pipes for such a distance has been proved at Pittsburg, where natural gas from the oil district is used in large quantities."

To give some idea of the vast importance of such a change Prof. W. Chandler Roberts, in a lecture at South Kensington in 1882, while expressing his belief that for domestic purposes we should ultimately adopt Sir William Siemens' plan of converting fuel into gas and burning it in a furnace quite separate from that in which the gas was produced, calculated that the soot that hung in a pall over London in a single day would be equivalent to at least fifty tons of coal, and added that there was good reason to fear that the carbon in the half-burned form of carbonic oxide gas was at least five times as much. He maintained that the presence of soot was always an indication of imperfect combustion, and therefore of waste.

The experiments of Sir William Siemens that ended in the perfection of the regenerative gas furnace, still considered capable of yielding such marvellous results, extended over a period of fourteen years; and during the next fourteen years he directed his inventive faculties to the production of steel by means of this furnace. His object was to make steel direct from the raw ores, without the intermediate use of huge blast furnaces and laborious refining processes. Other metallurgists had laboured to attain the same end, but had failed, partly on account of the want of such a furnace as he had now brought into successful operation.

When, in 1861, he took out his patent for that furnace, he stated that it was specially applicable to the melting of steel on the open hearth; and in the same year he suggested to Mr. Abraham Darby, of Ebbw Vale, that it should be used for the production of steel upon an open hearth; but no attempt was then made to put the idea to the test of practical experiment. He designed an open-hearth furnace in 1862 for a Durham iron-maker who was trying to make steel by melting a mixture of wrought iron and spiegeleisen; but it was found to be imperfect. Next year a furnace was made from his designs at Montluçon, in France, and a long series of experiments were made there. Good steel was produced; but unfortunately the heat of the furnace was inadvertently carried so high that the roof was melted, and the proprietors becoming unnerved, the process was abandoned. Two of his first licensees—a large Durham ironmaster and the then Inspector-General of Mines in France—succeeded in 1865 and 1866 in producing steel upon the open hearth; but they did not persevere sufficiently to make the process a commercial success. In 1866 he designed a furnace for a Glasgow firm, but it was also abandoned after a few days' trial. In the following year the Barrow Steel Company tried to make steel by his open-hearth process, and good steel was produced; but it was not sufficiently profitable to be continued. He now perceived that it was necessary for himself to solve the various difficulties which others regarded as practically insuperable. Having, he says, been so often disappointed by the indifference of manufacturers and the antagonism of their workmen, "I determined, in 1865, to erect experimental or 'sample steel works' of my own at Birmingham, for the purpose of maturing the details of these processes before inviting manufacturers to adopt them. The first furnace erected at these works was one for melting the higher qualities of steel in closed pots, and contained sixteen pots of the usual capacity. The second, erected in 1867, was an open-hearth furnace, capable of melting a charge of twenty-four cwt. of steel every six hours. Although these works have been carried on under every disadvantage, inasmuch as I had to educate a set of men capable of managing steel furnaces, the result has been most beneficial in affording me an opportunity of working out the details of processes for producing cast steel from scrap iron of ordinary quality, and also directly from the ore, and in proving these results to others."

His labours were now attended with more encouraging results. In 1867 he sent several samples of steel produced by his own process to the Universal Exhibition at Paris, and was there awarded a grand prize for his regenerative furnace and steel process. Mr. Ramsbottom, the engineer of the London and North Western Railway, examined the process in operation at Birmingham in a more advanced stage of development than had hitherto been reached ; and early in 1868 he adopted it at Crewe, where it was accordingly first worked on a manufacturing scale. About the same time, the directors of the Great Western Railway, having heard that by this process old iron rails could be converted into steel, in May of 1867 sent a truck load of old iron rails, which had originally been made at Dowlais, to the Sample Steel Works at Birmingham to be remanufactured into steel. Sir William Siemens, not without some unsuccessful experiments, eventually succeeded in converting some of them into steel, which was rolled into rails by Sir John Brown and Company of Sheffield. These rails, which the railway directors laid down at Paddington in the same year, though subjected to more than ordinary wear, were not taken up till 1878, and then they were not worn out, but were removed because the flanges of the carriage wheels had struck the bolts. So satisfied were the directors with the success of that experiment that the Landore Siemens Steel Company—the largest of its kind—

was immediately formed under the direction of Mr. L. L. Dillwyn, M.P. About the same time Messrs. Martin, of Sireuil, having obtained a license from Sir William Siemens, succeeded in making steel in the regenerative gas furnace by melting wrought iron and steel scrap in a bath of pig metal. The Siemens process produced steel from pig metal and iron ores. The former became known as the scrap process, and the latter as the direct process.

In 1869 these new processes were being carried out upon a large scale in England at the works of the Landore Siemens Steel Company, where seventy-five tons of steel per week were then being produced, at the works of the Yorkshire Steel and Iron Company, the Bolton Steel and Iron Company, the London and North Western Railway Company; and on the Continent at the works of Messrs. Verdie and Cie., Messrs. De Wendle and Cie., the Sireuil Steel Company, Messrs. F. Krupp and Company of Essen, Chevalier Stummer of Vienna, and others. Some thousands of tons of first-class steel manufactured by these processes were sold in the market in 1869.

The Siemens process, which has since then been more largely used every year, is thus described by the experienced manager of one of the largest works established for working it—Mr. J. Riley, manager for the Steel Company of Scotland: " The charge consists mainly or entirely of pig-iron, which is placed on the bottom and round the sides of the furnace. Melting requires four or five hours; then ore of pure character is charged cold into the bath, at first in quantities of four to five cwt. at a time. Immediately this is done a violent ebullition takes place; and when this has abated, a new supply of ore is thrown in—the object being to keep up uniform ebullition. Care is taken that the temperature of the furnace is maintained so as to keep the bath of metal and slag sufficiently fluid; but after the lapse of some time, when the ore is thoroughly heated, and reduction is taking place rapidly, the gas may be in part shut off the furnace, the combustion of the carbon in the bath itself keeping up the temperature. In the course of the operation, the quantity of ore charged is gradually reduced, and samples are taken from time to time of both slag and metal; when these are satisfactory, spiegeleisen or ferro-manganese are added, and the charge is cast. This mode of working has this advantage, that there is greater certainty as to the result, because of the known composition of the materials charged, which cannot be the case in dealing with large quantities of scrap, obtained, it may be, from a thousand sources."

Not content with the great success of this process, Sir William Siemens continued his experiments in the hope of inventing a still more direct process. Speaking on this subject in 1873, he said : " However satisfactory the results may appear that were obtained at the Landore and other works where the process just described is carried out upon a large scale in the production of steel for a great variety of purposes, I have never considered them in the light of final achievements. On the contrary, I have always looked upon the direct conversion of iron and steel from the ore without the intervention of blast furnaces and the refinery as the great object to be attained, and have in designing works made such provisions that the existing plant could be easily extended for the carrying out of a more perfect process. In proposing to produce iron and steel direct from the ore without the intervention of the blast furnace, I am aware that I shall be met by the objection that the direct process was practised by the ancient Indians and Romans, and had to give way to the blast furnace as the apparatus above all others well calculated to deal with ores in large masses, and which has in recent times been brought up to a degree of practical efficiency. Notwithstanding these facts, I do not despair of being able to prove that upon theoretical as well as practical grounds the blast furnace leaves much to be desired, and that the direct process, if dealt with according to our present state of chemical knowledge and mechanical resources, may be carried into effect with great practical advantages as regards economy of fuel, saving of labour, and quantity of material produced."

His first experiments worth mentioning, for this purpose, were made in a rotatory furnace which he erected at Landore in 1869, and which he described as consisting of a long cylindrical tube of iron, of about eight feet diameter, mounted upon anti-friction rollers, and provided with longitudinal passages in its brick lining for heating currents of air and gas prior to their combustion at the one extremity of the rotating chamber. The flame produced passed thence to the opposite or chimney end, where a mixture of crushed ore and carbonaceous material was introduced. By the slow rotation of this furnace the mixture advanced continually to the hotter end of the chamber, and was gradually reduced to spongy iron. This dropped through a passage constructed of refractory material on to the hearth of a steel melting furnace where a bath of fluid pig metal had been provided. The supply of reduced ore was so long continued that the carbon in the mixture was reduced to the minimum point. The rotation was then arrested to prevent further descent of the reduced ore; spiegel was added, and the contents of the melting furnace tapped into a ladle, and thence into ingots of steel. The reduction of the ore to the metallic state in this furnace was accomplished in a comparatively short time; and in that respect the process was successful. But the metal produced in it was found to contain an unusual percentage of sulphur, which it had absorbed from the heating gases, and which rendered it unfit for conversion into steel. The apparatus was consequently abandoned.

He next tried to accomplish his object in a reverberatory gas furnace, which he designed and put in operation; but its successful working depended, to a certain extent, on manual labour and skill. It then became evident, he says, that if iron and steel were to be produced largely by direct process, that

M

process must be a self-acting or mechanical one; and in 1870-71 his attention again reverted to the rotating furnace. He felt confident that if he could succeed in furnishing it with a lining capable of resisting the high degree of heat requisite for the precipitation of iron, and at the same time capable of resisting the chemical action, this mode of conducting the process must succeed. After many experiments he found that bauxite (from Baux in France) was the most suitable material for such a lining, for when exposed to intense heat it was converted into a solid mass of emery, of such extreme hardness that it could hardly be touched by steel tools.

With this lining and other improvements, the new rotatory furnace consisted of four regenerators of the ordinary kind with reversing valves and gas producers. It was so mounted that it could be made to rotate very slowly, or at the rate of one revolution per minute. At the one end of this rotating cylinder, on the same side as the regenerators, was an opening for the admission of the heated gases and air, as well as an outlet for the products of combustion, these two passages being separated by a vertical partition. At the other end of the cylindrical furnace was a door hung in the usual manner, and there was also a taphole on the working side for discharging the slag. The working chamber was heated very perfectly by the gases, which, entering with a certain velocity, traversed it to and fro, and then escaped by the exit passage. In this heated chamber while slowly revolving was placed ore broken up into small pieces, with the requisite lime or other material for fluxing. In about forty minutes this charge was heated to bright redness, and then small coal, in the proportion of about one-fourth of the charge of ore, was added. The rotative velocity being at the same time increased in order to accelerate the mixture of the coal and ore, a rapid reaction followed; metallic iron was precipitated by each piece of carbon; while the fluxing material formed a slag with the refuse from the ore. Thereafter the rotation was made slower, so that the mass inside might be turned over and over, presenting continually new surfaces to the heated lining and to the volume of flame, while only heated air, not gas from the gas producers, was admitted. When in this way the reduction of the ore was nearly completed, the fluid cinder was tapped off and the quick speed again resumed in order thereby to collect the loose masses of iron into two or three balls, which, being taken out, were either shingled or rolled in the usual way of consolidating puddled balls, or transferred to the bath of a steel-melting furnace, where they could be at once converted into cast steel.

Encouraged by the results of this process worked on a small scale at Birmingham in 1873, he ventured along with others upon some larger applications of it, principally at Towcester in Northamptonshire. There he soon discovered that the ore, so abundant in that country, although capable of yielding iron of good quality, was too poor and irregular to afford satisfactory commercial results, unless it was mixed with an equal weight of rich ore, which, as well as fuel, was expensive at Towcester owing to the high rates of carriage. However, three rotating furnaces were built there, and bars of iron produced in them were sold in Staffordshire and Sheffield at exceptionally high prices, being deemed equal to Swedish bar in toughness and purity. Iron and steel of very high quality were thus produced by direct process from the poorest ores, but at a cost which was found unremunerative. From a commercial point of view, therefore, it was unsuccessful, but so far from abandoning his purpose he renewed his experiments at Landore in 1880 with the rotating furnace, in which he had succeeded in making some further improvements.

While these experiments were going on, several attempts were being made by other metallurgists to invent or discover a profitable and effectual means of extracting pure iron from ironsand, which was found in large quantities in countries hitherto considered poor in minerals. Attention was being called to large beds of ironsand in New Zealand and in Canada, near the shores of the St. Lawrence River, containing 50 per cent. of metallic iron. Large sums of money had been spent in trying to free the iron from its impurities, but without success. In 1883 it was demonstrated that the work could be done successfully and rapidly by the direct process of Sir William Siemens. In his rotating furnace magnetic ironsand from Canada was easily reduced in less than four hours to iron balls, which were taken straight to the open-hearth furnace and converted into mild steel. The results were excellent. About the same time the process was successfully used at Pittsburg—the chief seat of the American iron trade; and, according to an account of its working given before the American Institute of Mining Engineers, it appeared capable of becoming a commercial success.

The invention of these two furnaces gives Sir William Siemens a unique position in the history of iron metallurgy. No other metallurgist has given the world new and easier processes for the complete conversion of raw ore into iron, and then into steel. The experiments which ended in this realisation of his views extended over a period of twenty-five years, and the rapid extension of the one process no doubt encouraged him to persevere with the other in the face of no ordinary difficulties and detraction.

The name of Emil Martin has often been associated with that of Sir William Siemens as an improver or contemporaneous inventor of the open-hearth process of steel making. Indeed the process is not unfrequently called the Siemens-Martin process. The way in which the name of Martin came to be connected with the process is not the least interesting episode in its history. At the Paris Exhibition of 1878 the Society for the Production of Martin Steel—the successors of Emil and Peter Martin—published a pamphlet in which they claimed that the credit of having invented and perfected the open hearth process belonged to Martin alone.

Sir William Siemens of course demanded a retractation, and issued a statement of his claims to priority. The Company declined to retract, whereupon Sir William published the entire correspondence in a pamphlet illustrated with drawings. In one of these letters he stated that he had worked at the solution of the problem of melting steel on an open hearth since 1856, and that in 1861 he took out a patent which was put into practical operation by Atwood in Durham, and by Messrs. Boignes, Rambourg, and Company of Montluçon in France. He called the attention of Messrs. Martin to the fact that on the occasion of his first negotiations with them, in a letter dated May 26, 1863, he informed them of the experiments at Montluçon. The correspondence showed that the question of the application of the Siemens regenerative system by the Martins referred only to the crucible steel furnace ; and that Sir William's idea of applying the open-hearth furnace to steel making was a novelty to the Martins, who, in accordance with his original request, agreed to the condition that the furnace was to be a reheating one, which might at a small expense be altered to an open-hearth steel furnace. The first furnace at Sireuil was begun, according to Sir William Siemens' plan, on April 17, 1863, and put in operation by his engineers. He acknowledged that after 1864 the Martins followed up the process with great perseverance, and that it was especially the proper mixture of the materials which they studied. It was only in 1867, after having concluded their experiments and begun regular work, that the Martins took out a new patent, in which many points of the open-hearth steel process were embodied, as they were also in the patent granted in the same year to Sir William Siemens. The Martin patent contained recipes for the production of cast steel capable of being hardened, of a homogeneous metal which would not harden, and of a " mixed metal "—a mixture of cast iron and steel. These recipes are now of no practical value.

If there were really any question as to the originality or priority of invention in this case, the opinion of English, German, and French metallurgists might be regarded as liable to bias; but the question has been discussed by a body of metallurgists that cannot be charged with national prejudice or partiality. The Mining and Metallurgical Association of Styria and Carinthia debated the point with more than ordinary information; and in the course of the debate Professor Kupelwieser stated that in his opinion neither Sir William Siemens nor the Martins "originated" the open-hearth process; and he called attention to the fact that Professor Gruner, in a paper published in the *Annales des Mines* in 1868, mentioned a description in an article in Hassenfratz's *Siderotechnik* for 1812 of a process used in an English ironworks of melting cast and wrought iron in a reverberatory furnace, sampling, then ladling, and casting. This process was taken up again, and in the years succeeding 1850 similar experiments were made by Colonel Alexander at Brest, but without result, owing to the poor quality of pig iron used. The process was not new, therefore, in France or in England; but there can be no doubt that it could not be successful at such temperatures as were produced by furnaces before Sir William Siemens' invention. Both Professor Kupelwieser and Director Sprung maintained that no royalties need be paid in Austria on the so-called Martin process; and the following resolutions were proposed by the chairman, Professor Tunner, and were passed unanimously:—

1st. The principle of the production of cast steel in a reverberatory furnace was known in England before the year 1812; and in 1860 Sudre, under orders from Napoleon III., carried it out successfully on an open hearth in the Montolaire works.

2nd. The idea of melting steel in the Siemens furnace originated with Sir William Siemens in the year 1862, and Martin built an open-hearth furnace which might be cheaply, according to

Sir William Siemens' directions, altered into a steel furnace. In April, 1863, Sir William Siemens' engineers built the first Siemens-Martin steel furnace at Martin's works at Sireuil.

3rd. In the year 1864 Martin discovered the proper additions for the various grades of steel, and received a patent for them on August 15, 1865. The furnace shown in the drawing of this patent was identical with Sir William Siemens' invention, and Sir William Siemens' drawing was also added to the subsequent patent of August 21, 1867.

4th. Martin can claim priority for his additions only.

5th. As these additions have been entirely superseded by the manipulations based upon more recent experience, Martin's patent is now of no value.

6th. Martin's claim has been rejected in France also, nobody there paying patent royalties.

Such, in short, is the way in which the name of the Martins became affiliated with the Siemens process; and as Sir William Siemens has handsomely expressed his appreciation of the labour and skill that they applied to the carrying out of the process in the early days of its existence, they have been amply rewarded for their pains. To speak of them as the inventors of the process would be playing with words. The account which we have given of Sir William Siemens' experiments in perfecting the direct process during a quarter of a century amply vindicates his claims; but it is proper to add that he has always shown a chivalrous sense of honour in giving others credit for any suggestion that may have been of use to him.

Sir William told the Parliamentary Committee on Patents that he would not have continued his long and costly experiments with the gas regenerator and open-hearth process if the English patent law had not insured such a period of protection as would repay him for his labour. As an illustration of the wisdom of his choice in thus coming to England to dispose of his inventions and of the reward which inventors meet with in their own country, it may be mentioned that he and his brother applied to the German Government for a patent for the regenerative furnace, and it was refused on the ground that in the Middle Ages stones were heated and thrown into cellars of town halls or other public buildings in order to warm them ; and that, forsooth, was considered a valid reason by the Fatherland for refusing a patent for one of the greatest inventions ever made by a German. In the House of Commons in 1883 the President of the Board of Trade mentioned the regenerative furnace as one of the most valuable inventions that had ever been produced under the protection of the English patent law, and stated that it was used by nearly every industry in the kingdom. In the hearing of an action in the superior courts of the United States in 1883 as to the duration of the patent in that country, it was stated that the inventor had received a million dollars in royalties. The Siemens steel furnace has also been a great success. Although it has had to compete with the Bessemer converter, it has made wonderful progress. Notwithstanding the commercial depression then prevailing in the iron trade, the production of open-hearth steel in the United Kingdom increased from 77,500 tons in 1873 to 436,000 in 1882. At the end of that year there were over 150 open-hearth furnaces in this country and more new ones were in course of construction. As in the Bessemer process, the ores then used in the open-hearth furnace were low in phosphorus ; but in 1880 it was found in the working of new furnaces at the Parkhead Works, Glasgow, that by employing an excessively high temperature phosphorus could be almost entirely eliminated in the open-hearth process. In the first charges of metal at these works the phosphorus stood at ·07, and in the first samples was reduced to ·008—or a reduction of 90 per cent.—a result which surprised Sir William Siemens, who stated that he had never before known such a reduction of phosphorus take place in the open hearth. The process has also rapidly extended on the Continent.

This process is considered best adapted for the production of large and heavy pieces of steel; and, being slower in its operation, the metal can be more easily tested and its quality more accurately regulated than in any other process. It is also exceptionally useful for converting old iron rails into steel.

In rolling steel armour-plates for the Admiralty the Landore Steel Company made the experiment of casting the ingot in a special mould, and after it had been hammered into a slab weighing three tons, it was rolled into a plate 3 in. thick and then planed to a size of 8 ft. by 3 ft. 6 in. The plate was next planed through the middle to test its soundness. The experiment was considered so satisfactory that the Admiralty in 1876 gave the Landore Steel Company orders to make all the plates required in the construction of Her Majesty's despatch vessels *Iris* and *Mercury*, of 3,735 tons displacement, which, with the exception of the rivets, were made entirely of Siemens steel. Thereafter the Admiralty determined to procure all their steel plates from the Landore works ; and so satisfactory were the results that no other material is now used in the Royal dockyards in the construction of the boilers and hulls of vessels. The decks of armour-plated ships are now also made of 2 in. plates of mild steel made by the open-hearth process.

It was also the special adaptability of this kind of steel for shipbuilding that reconciled shipbuilders to its use in mercantile vessels. Notwithstanding that, in this instance, the British Admiralty led the way, mercantile owners were very slow to patronise the "new metal." The manager of the works of the Steel Company of Scotland, Mr. J. Riley, who was in advance of his contemporaries in this matter, says he well remembers the anxiety he experienced in 1877 to obtain a contract for the supply of steel for a merchant vessel. This was at last accomplished at

Newcastle. The ice was thus broken, but progress was disappointingly slow. The Admiralty followed up their first venture by contracting for six corvettes to be built of steel by Messrs. Elder and Co. The owner of the merchant vessel, which Mr. J. Riley's energy and foresight initiated, had such satisfactory returns from her, that he contracted for a second vessel. Meanwhile, the use of steel had been engrossing more and more of the consideration of shipowners and builders. Many small ventures were made, until at length the great companies determined to try steel instead of iron. Messrs. Allan led the way with the *Buenos Ayrean*, and were quickly followed by the Pacific Steam Navigation Company, Sir Donald Currie, the Peninsular and Oriental Company, the British Indian Company, the White Star Company, the Cunard Company, the Orient Company, and many others. One of the grandest of the early vessels built of steel was the *Servia*, which was lauched on the Clyde in March, 1881. This was then the largest addition to the Cunard line, and was considered one of the finest specimens of marine architecture. Besides the special attention paid in her construction to insure the comfort and safety of ordinary passengers, her arrangements complied with the Admiralty requirements that qualify her for purposes of war. It has been found by experiment that eight feet of coal will stop a shot from a sixty-four pounder, and that a shell will explode harmlessly in far less. The engines and boilers of the *Servia* are surrounded by watertight compartments, available for the reception of coal ten feet in thickness. This would give complete protection to the vital parts of the ship. Besides this, the whole length of the vessel is divided by nine watertight bulkheads, each of which is to all intents and purposes a separate ship, provided the doors, which can be closed by a lever on the main deck, are shut in time of danger. The ship is not only built of steel, but has a double skin, so that, were the outer plates broken, the ship would still be safe. The upper, main, and lower decks are also of steel, covered with pine on the upper deck and teak below. The *Servia* has capacity for storing sufficient coal to enable her to cruise for two or three months at a fair speed without entering a port, and she can go at the remarkable rate of twenty and a half statute miles an hour with an almost total absence of vibration.

Increasing experience proved incontestably that vessels built of mild steel are much safer than those built of iron ; there is less risk of loss with them ; and, being lighter than iron ones, their earning power is so much increased that they make very handsome returns for the additional first cost of steel. In the case of the steel vessels already referred to, the increase of income, as compared with iron vessels, was 25 per cent.

Instances of the superior power of steel vessels to withstand shocks were not uncommon. The steel plates of the *G.M.B.*, a vessel belonging to Messrs. James Watson and Co., of Glasgow, after she had been in collision at sea, were bent and crumpled up in a fearful style, yet not one was cracked. It was the opinion of experts who saw her after the collision that had she been built of iron she must have inevitably sunk. Another instance was related by Mr. William Denny, of Dumbarton, to the Institute of Naval Architects :—" The *Rotomahana* had a very narrow escape from total loss. She was engaged in an excursion from Auckland to Great Barrier Island, a distance of fifty miles, and was leaving the harbour of Fitzroy (in Great Barrier Island) by a somewhat difficult passage, when she struck on a sunken rock with considerable force. She made some water on the way back to Auckland, as it afterwards turned out, through some rivet-holes ; these were plugged, and she was enabled to return to Dunedin to be docked. The worst damaged plate was taken out, re-rolled, and replaced. Several frames were set back, and a good job made of the repairs within seventy-two hours. This experience has shown clearly the immense superiority of steel over iron.

There is little doubt that, had the *Rotomahana* been of iron, such a rent would have been made in her that she would have filled in a few minutes. A number of frames were set back by the force of the blow; the bulkhead was bulged, and the plate was corrugated, and yet there did not appear one crack anywhere."

In these circumstance the use of steel for shipbuilding increased with unexampled rapidity. In 1879, only about 20,000 tons of steel vessels were built, while in 1883 over 260,000 tons were built, being one-fourth of the total tonnage of new shipbuilding for that year.

CHAPTER VII.

" The inheritance of great men is their invention : their heirs are the human race."—LAMARTINE.

THE names of the Brothers Siemens will for ever be associated with the application of electricity to industrial purposes. We have already recorded the first notable incident in connection with Sir William Siemens' discoveries in electrical science at a time when he was in his minority and electricity was in its infancy. At that time Faraday was the high priest of that science, and his mantle may be said to have fallen on Sir William Siemens. In 1821 Faraday thought he would be rendering a service to himself and others by writing a history of electro-magnetism; and the final issue of his studies was the discovery in 1831 of a new domain of electricity, called by him magneto-electricity.

To illustrate this principle, which Sir William Siemens was the first in this country to utilise on a large scale, Professor Tyndall says : Take two flat coils, and conceive one of them to have its two ends united together, so as to form a complete circuit; conceive no current to be flowing through one coil ; imagine then a current from an electric battery sent through the other coil. If you suppose one coil to be held above the other, with a current flowing through the former and none through the latter, and you cause one coil to approach the other, simply by that approach you evoke in the coil which is not at all connected with the battery an electric current. That electric current is only generated during the motion of one coil towards the other, and the moment that motion ceases the current ceases. If, having produced this current by the approach of the two coils, you afterwards separate them, you get another current opposite in direction to the first. Thus by approach and retreat you get currents in opposite directions ; and to these particular currents Faraday gave the name of "induced currents." Again, Faraday laid one coil upon the other, a current flowing through neither ; then he started a current through one coil, and instantly, by a kind of reaction or kick, a momentary current was evoked in the other coil not connected with the battery. Interrupting the battery current, the subsidence of the current in the one coil caused a current in the other, opposite in direction to the first. Then Faraday passed on to examine the part played by magnetism in the production of these currents. Supposing that you have a piece of iron in a coil, you cannot alter the magnetic condition of that piece of iron to the slightest extent without evoking in the surrounding coil one of Faraday's induced currents. In illustrating that point Faraday brought a magnet up to the end of such a bar of soft iron, and instantly he saw his needle start aside through the production of the induced current, due to the magnetisation of the piece of iron. When he withdrew his magnet, the magnetism of the iron subsided. The law is that the current produced by the exaltation of the magnetism is always opposed in direction to the current evoked by the subsidence of the magnetism.

Those who have seen the noble statue of Faraday by Foley will have observed something that looks like a garland in Faraday's hand. It is not a garland ; it is a representation of an iron ring which was constructed by Faraday himself. That iron ring he covered by two distinct coils entirely separate from each other. By sending a current from a voltaic battery through one of these coils, Faraday magnetised the ring ; and by the magnetisation he obtained his induced current in the other coil. That ring remained in the possession of Sir James South for a great many years, and soon after his death it came into the possession of the Royal Institution, being the very first ring used by Faraday himself in illustration of this subject.

Though Faraday showed that permanent magnetism might be made to produce electric currents, and that electric currents also generate electric currents, he did not apply his discoveries to the practical purposes of producing electric heat, or electric light, upon an economical scale ; but, with the prescience of a true man of science, he predicted that his results, which were exceedingly feeble in the first instance, would receive their full development hereafter.

In lecturing on this subject before the Institution of Civil Engineers in 1883, Sir William Siemens exhibited in operation the original instrument by which Faraday had elicited the first electric spark before the members of the Royal Institution in 1831, explaining that, although the individual current produced by magneto-induction was exceedingly small and momentary in action, it was capable of unlimited multiplication by mechanical arrangements of a simple kind, and that by such multiplication the powerful effects of the dynamo-machine of the present day were built up. One of the means for accomplishing such multiplication was the Siemens armature of 1856. It consisted of a piece of iron with wire wound round it longitudinally, not transversely ; and was the most powerful and perfect apparatus of its kind.

Ten years afterwards Sir William Siemens in London and Dr. Werner Siemens in Berlin were the first to announce an application of this armature that may eventually prove of as great practical importance in the development of electricity as Watt's engine was in the application of steam. On the 4th of February, 1867, Sir William Siemens sent to the Royal Society a paper " On the Conversion of Dynamic into Electrical Force without the Use of Permanent Magnetism." Ten days afterwards the Royal Society received a paper from Sir Charles Wheatstone bearing the title. "On the Augmentation of the Power of a Magnet by the Reaction thereon of Currents induced by the Magnet itself." Both papers announced the same discovery, and were illustrated by experiments. Both were read upon the same night—the 14th of February. "It would be difficult," says Professor Tyndall, "to find in the whole field of science a more beautiful example of the interaction of natural forces than that set forth in these two papers. You can hardly find a bit of iron—you can hardly pick up an old horse-shoe, for example—that does not possess a trace of permanent magnetism; and from such beginnings Siemens and Wheatstone have taught us to rise by a series of interactions between magnet and armature to a magnetic intensity previously unapproached." In its simplest form the mechanism consists of a plate of iron bent into a horse-shoe form and coiled round with insulated copper wire; and between the ends of it rotates a Siemens armature. The wire from the armature is connected with the wire passing round the bent piece of sheet iron, which is called the electro-magnet. When the handles are turned, in the first instance there are induced currents of infinitesimal strength produced in the armature which rotates between the poles of the electro-magnet. "Instead of trying to utilise these infinitesimal currents, they are carried round the magnet, and the magnet's power is thereby exalted. The exalted power is brought immediately to bear upon the armature, producing in it also currents of exalted strength. Those currents are again sent round the electro-magnet, which has its power enhanced by them. The electro-magnet, with its power thus enhanced, reacts again upon the armature; and thus, by a play of mutual give and take between the armature and the magnet, that magnet is raised from infinitesimal strength to a state of magnetic saturation. Although no part of the rotating armature touches the bars which are excited by the current produced, yet the power accumulates, or the resistance increases, with the velocity to an extent limited only by the ultimate power of the iron to become magnetic."

A suggestion, contained in Sir Charles Wheatstone's paper, that "a very remarkable increase of all the effects, accompanied by a diminution in the resistance of the machine, is observed when a cross wire is placed so as to divert a great portion of the current from the electro-magnet," led Sir William Siemens to an investigation which was described before the Royal Society on March 4th, 1880, and in which it was shown that by augmenting the resistance upon the electro-magnets a hundredfold, valuable effects could be realised. The most important of these results consisted in this, that the electro-motive force produced in a "shunt-wound machine," as it was called, increased with the external resistance, whereby the great fluctuations formerly inseparable from electric-arc lighting could be obviated, and that, by the double means of exciting the electro-magnets, still greater uniformity of current was attainable.

It was the invention of the dynamo machine that made practicable the application of electricity to industrial purposes. Experiments have shown that it is capable of transforming into electrical work 90 per cent. of the mechanical energy employed as motive power. Sir William Siemens has himself described it as perhaps the most beautiful illustration of the convertibility of one form of energy into another. To attempt a description of its usefulness would be like trying to answer Franklin's question, "What is the use of a new-born child?" Its practical application is still in its infancy. It was in 1785 that Watt finished his "improvements" in the steam-engine; and the century that has since elapsed has not sufficed to demonstrate the full extent of its utility. The next hundred years will probably witness a similar extension of the dynamo machine to practical purposes. It is yearly giving fresh evidences of its utility.

It was the absence of sufficient electrical power that delayed the

N

use of the electric light for seventy years. Davy produced an electric light in 1808 ; but its cost was so great that some members of the Royal Institution had to subscribe liberally to defray the expense of the experiment. Strength of current was what was wanted. Given a sufficient electrical current, it was known, according to Professor Tyndall, that the next condition to be fulfilled in the development of light and heat was that it should encounter and overcome resistance. "A rod of unresisting copper carries away uninjured and unwarmed an atmospheric discharge competent to shiver to splinters a resisting oak. Send the self same current through a wire composed of alternate lengths of silver and platinum ; the silver offers little resistance, the platinum offers much. The consequence is that the platinum is raised to a white-heat, while the silver is not visibly warmed. The same holds good with regard to the carbon terminals employed for the production of the electric light. The interval between the terminals offers a powerful resistance to the passage of the current, and it is by the gathering of the force necessary to burst across this interval, that the electric current is able to throw the carbon into that state of violent intestine commotion which we call heat, and to which its effulgence is due."

In the development of the appliances for the production of this light Sir William Siemens has taken a leading part. But while ever zealous to promote its progress, he has never taken a partisan view of its utility. He candidly admits that gas will continue to be the poor man's friend. In 1882 he told the Society of Arts that "electricity must win the day *as the light of luxury,* but gas will find an ever-increasing application for the more humble purposes of diffusing light." He estimated the cost of lighting the whole of London by electricity at £14,000,000, exclusive of lamps and internal fittings, and the cost of extending the same to the towns of Great Britain and Ireland was calculated at £80,000,000.

But the electric light is only one of several useful purposes for which the dynamo machine has been utilised by Sir William Siemens. In June, 1880, he electrified the Society of Telegraph Engineers by exhibiting the power of an electrical furnace designed by him to melt considerable quantities of such excessively refractory metals as platinum, iridium, and steel. He explained that he was led to undertake experiments with this end in view, by the consideration that a good steam-engine converts 15 per cent. of the energy residing in coal into mechanical effect, while a good dynamo-electric machine is capable of converting 80 per cent. of the mechanical energy into electric energy. If the latter could be expended without loss within an electric furnace, it would doubt-less far exceed in economy that of the air-furnaces still largely used in Sheffield. In the small furnace which he exhibited before the telegraph engineers, the positive electrode, made of iron, entered from below the crucible containing the metal to be melted, while the negative electrode—a rod of carbon—was attached by means of a lever to a solenoid regulator. The crucible was sur-rounded by charcoal contained in a copper vessel to prevent loss of heat, and so intense was the heat accumulated that in about twenty minutes two pounds of broken files were completely melted. He showed that the apparatus was one that could be easily applied on a large scale.

He may also be fairly described as the creator of electro-horticulture. Some experiments that he made early in 1880 led him to the conclusion that the electric light could produce the colouring matter in the leaves of plants, and promote the ripening of fruit at all seasons of the year and at all hours of the day or night. He found that plants do not require a period of rest during the twenty-four hours, but make increased and vigorous progress if subjected during the day to sunlight and to electric light at night. These observations on combined sun and electric light agreed with those made by Dr. Schübeler, of Christiania, who found as the result of continued experiment in the north of Europe during an Arctic summer that plants, when thus continuously growing, develop more brilliant flowers and larger and more aromatic fruit than when under the alternating influence of light and darkness.

In the winter of 1880 he put the conclusions he had thus arrived at to the test of experience on a large scale at his country residence near Tunbridge Wells; and the results obtained were communicated to the British Association at York in 1881. The use of the electric light in a variety of ways proved that it most effectually promoted vegetation when it was surrounded by a clear glass lantern. Under these conditions he stated that " Peas, which had been sown at the end of October, produced a harvest of ripe fruit on the 16th of February, under the influence, with the exception of Sunday nights, of continuous light. Raspberry stalks put into the house on the 16th of December produced ripe fruit on the 1st of March, and strawberry plants put in about the same time produced ripe fruit of excellent flavour and colour on the 14th of February. Vines which broke on the 26th of December produced ripe grapes of stronger flavour than usual on the 10th of March. Wheat, barley, and oats shot up with extraordinary rapidity under the influence of continuous light, but did not arrive at maturity; their growth, having been too rapid for their strength, caused them to fall to the ground after having attained the height of about twelve inches. However, seeds of wheat, barley, and oats planted in the open air and grown under the influence of the external electric light produced more satisfactory results; having been sown in rows on the 6th of January, they germinated with difficulty on account of frost and snow on the ground, but developed rapidly when milder weather set in, and showed ripe grain by the end of June, having been aided in their growth by the electric light until the beginning of May. Doubts have been expressed by some botanists whether plants grown and brought to maturity under the influence of continuous light would produce fruit capable of reproduction ; and in order to test this question, the peas gathered on the 16th of February, from the plants which had been grown under almost continuous light action, were replanted on the 18th of February. They vegetated in a few days, showing every appearance of healthy growth." Mr. Darwin and other authorities were previously of opinion that many plants, if not all of them, required diurnal rest for their normal development; but these experiments in electro-horticulture led Sir William Siemens to the conclusion that, although periodic darkness evidently favours growth in the sense of elongating the stalks of plants, the continuous stimulus of light was favourable to healthy development at a greatly accelerated pace, through all the stages of the annual life of the plant, from the early leaf to the ripened fruit. The latter was superior in size, in aroma, and in colour to that produced by alternating light. The beneficial influence of the electric light was very manifest upon a banana palm, which at two periods of its existence—viz., during its early growth and at the time of the fruit development,—was placed (in February and March of 1880 and 1881) under the night action of the electric light, set behind glass at a distance not exceeding two yards from the plant. The result was a bunch of fruit weighing 75 lbs., each banana being of unusual size, and pronounced by competent judges to be unsurpassed in flavour. Melons also remarkable for size and aromatic flavour were produced under the influence of continuous light in the early spring of 1880 and 1881. In conclusion, he expressed his belief that the time is not far distant when the electric light will be found a valuable adjunct to the means at the disposal of the horticulturist in making him really independent of climate and season, and furnishing him with a power of producing new varieties, while the electric transmission of power may eventually be applied to thrashing, reaping, and ploughing.

Sir William Siemens has also been a pioneer in the introduction into this country of the electric railway, which was originally invented by his brother, Dr. Werner Siemens. Like most novelties, it was not brought into useful operation without encountering opposition.

The first electric railway was shown at the German Industrial Exhibition in 1879. It was from 400 to 500 yards in length. The electric locomotive was one of four and a half horse power, and it drew a train of miniature cars for eighteen or twenty persons. While this was being shown Dr. Werner Siemens proposed to erect an elevated electric railway at the expense of his firm in the fashionable Frederic Street in Berlin; but the inhabitants being opposed to it, the Emperor vetoed it before he was asked to sanction it. He next proposed to erect elevated electrical railways in some of the busiest parts of the city, in order to relieve the streets of a great deal of their crowded traffic. The police authorities opposed this project, but indicated that it might be convenient to construct such a railway in some other quarter. A fresh plan for a general network of railways for the city was submitted to the authorities, who were asked to select the quarter which they thought the most convenient. Two months afterward the authorities replied that Berlin was not in want of electrical railways. The resolute inventor next proposed to have a railway in the suburbs of Berlin to connect the Lichterfelde Station of the Anhalt Railway with the Central Military School, a distance of nearly two miles; and after some legal difficulties were got over the local authorities approved of the execution of this scheme. The work of construction was soon finished. The inventor wished to make it an elevated railway, but he was deterred from doing so by the expense; it was therefore put upon the ground, though an elevated railway was always advocated by him as the most economical. The electrical locomotive not being heavy, like a steam-engine, the permanent way can be made lighter than usual ; and the elevated line affords the requisite means of insulation. In the Lichterfelde line nothing was done to effect insulation from the earth, and consequently a great loss of the electrical current was expected. Nevertheless it worked very successfully. When opened in May, 1881, the train consisted of a carriage, constructed to carry twenty persons, and the four and a half horse electro-motive engine, which was considered capable of running from twenty to twenty-two miles an hour. The maximum speed permissible, however, was under twelve miles an hour. This was accomplished with ease even over a part of the line where the gradient was one in a hundred. Fifteen journeys were made daily in connection with the train service of the Anhalt Railway ; but the afternoon traffic being considerable, additional journeys were made to carry the excess of passengers to the main line station, to which the electrical railway formed a tributary. The only accident that occurred during the first six months' working was the startling of two horses that received a shock from the electric current on the rails at one of the street crossings. At this the local authorities took alarm, but this source of apprehension was removed by the easy and perfect expedient of insulating the portion of the line forming the crossing and allowing the train to run over it by its own impetus, the electrical current being carried by a copper wire under the rails, and picked up again by the electrical locomotive after the crossing had been effected. This railway, which was made at the sole cost of the Messrs. Siemens and worked by them, carried sufficient freight to fully cover its working expenses.

Dr. Werner Siemens next determined to work by his electric system an existing line of ordinary tramway a mile long between Charlottenburg, his own estate, and the Spandauer Bock. As it was impossible there to insulate the rails, he made experiments in the use of overhead wires in order to keep the electrical current clear of the line. Overhead wires were mounted on telegraph poles placed at the side of a line of tramway in the Berlin works ; and the ordinary tram-cars were run upon that line by means of conductors attached to pulleys running upon the overhead wires. This experiment proving successful, the Messrs. Siemens were preparing the designs for its application to the Charlottenburg tramway, when they were asked to exhibit their locomotive railway at the Paris Exhibition of Electricity ; and they readily consented to do so upon their new principle. Accordingly, a tramway was specially constructed by them from the Place de la Concorde to a station within the Palace de l'Industrie, a distance of half a mile. It was opened on the 31st of August, 1881, and the King of the Sandwich Islands was a passenger on one of the trial trips. The tram-car was of the same pattern and dimensions as that in use on the ordinary tramways in Paris. It carried forty-six passengers, and the speed varied from eight to fifteen miles an hour. The driving power came from a fixed engine in the Exhibition turning a Siemens' generator, the current from which was conveyed to two metal rods, carried on posts parallel to the tramway, about 10 feet from the ground. Little rollers running on these were connected by wires with a second electric machine in the base of the tram-car and geared to its wheels. The energy of the stationary steam-engine was conveyed along the wire in the form of an electric current, and, being reconverted into mechanical energy in the second machine, turned the wheels of the tram-car and propelled it. The little rollers were drawn along the wire as the car moved, and kept up a continuous electrical connection. It was a great success. During the first month it was in operation it made 21,000 journeys, and carried 50,000 passengers. The total distance travelled in that time was equal to the distance between Paris and Berlin, and although the ground over which it travelled was crowded with the ordinary traffic of the town, it was worked without any accident.

The first practical application of electricity for tramway or railway propulsion in the United Kingdom was on a new line of tramway between the Giant's Causeway and Portrush, a distance of about six miles, where the motive power could be obtained from a neighbouring waterfall, hitherto unutilised. At the commencement of this undertaking, in the last week of September, 1881, the chairman of the company, Dr. Traill, informed the directors and a large gathering of local gentry that they had assembled not merely to inaugurate an obscure or local work, but to introduce into Ireland for the first time one of those scientific discoveries in which the last quarter of a century had been so fruitful. This tramway would be worked by electricity, and under the direct auspices of Sir William Siemens, who was a member of the Board and a large contributor to the funds. It had always been remarkable, he said, that the most brilliant scientific discoveries appeared, when known, to be the simplest. Most of the properties of magnets and electric currents had now been known for a long time, but it was only quite recently that electricity and magnetism had come to be applied to locomotive purposes, and, so far, with such success as to justify the prediction that they were to find in them the great motive power of the future. Not many years would elapse before this dynamo-electric power would be supplied, not alone to tramways suitably situated for it, as this one undoubtedly was, but also to railways. Shareholders in a company such as this could easily see what an important thing such a revolution in locomotive power would represent. The working expenses for haulage on a tramway such as theirs with horses would be about 11*d.* per mile, and by steam power about 7*d.* per mile, but there was every reason to suppose that the working expenses of their motive power need not reach 1*d.* a mile. Further, as each car would carry its own locomotive power, they would save the expense of engine-drivers and stokers, and all that class of persons, as well as effect an immense saving in fuel; and, what was more important, as they required no heavy engines to increase the friction and to take a grip of the rails for hauling purposes, their rails would not suffer great wear and tear, but would only have to bear the weight of the traffic in light cars. The Provost of Dublin University, the Rev. Dr. Jellett, said that as a scientific man, and as the head of a great scientific institution, he took much more than a merely local view of the enterprise just commenced. It was the inauguration of a new era of locomotive power in these islands. Most scientific men knew that the most difficult thing they had to encounter was to correct or economise force; and in this scheme they were about to utilise in a new way the large forces of nature, which were at present going to waste.

The new line was opened by the Lord Lieutenant of Ireland (Earl Spencer), on September 28, 1883.

Respecting the future development and use of this system, Sir William Siemens has repeatedly stated that though the experience gained in the working of the first electric railway at Lichterfelde left no reasonable doubt regarding the economy and certainty of this mode of propulsion, he did not anticipate that it would supersede locomotive power upon our main trunk railways. "It will have plenty of scope in relieving the toiling horses on our tramways, in use on elevated railways in populous districts, and in such cases as the Metropolitan Railway, where the emission of the products of combustion causes not only the propulsion, but the suffocation of passengers."

According to the *Revue Industrielle* of July 19, 1882, there were then about 100 miles of electric railways working, authorised, or in course of construction; and grants for their construction were becoming more numerous. Lines were being projected in Germany, Austria, Holland, Italy, and the United States.

Though the novelty of the latest applications of electricity appears to have almost put the telegraph in the shade, yet the services of the Brothers Siemens in bringing it to perfection have been not less original and useful. The University at which Sir William Siemens finished his education was the cradle, if not the birthplace, of the electric telegraph. According to Sir William's own account of it, the celebrated astronomers and physicists of Göttingen—Gauss and Weber—established in 1833 a line wire reaching from the observatory of that University to the steeple of the public library, and thence to the magnetic observatory—a distance of about a mile —a return circuit being also provided. Through this circuit they communicated with each other by means of magneto-electric currents and a Weber's reflecting magnetometer, and notwithstanding the large proportions of this receiving instrument, a needle weighing nearly one hundredweight, they succeeded in obtaining very clearly defined signals. Being themselves engaged in scientific pursuits, they called upon Steinheil, of Munich, to construct a practical and useful electric telegraph. Steinheil applied himself vigorously to the task, and produced a telegraphic system which would have been nearly perfect if it had not been too refined for the means then at his disposal. Sir C. Wheatstone, of London, and Mr. Morse, of the United States, were simultaneously working at the same problem, and each claimed the honour of having solved it. The telegraph, however, was still in an infantine state when the Brothers Siemens began to study it, and their series of inventions largely aided in bringing it to perfection. A description of all the various mechanical appliances which they produced would be out of place here ; but some of their greater achievements, which form epochs in the history of telegraphy, are of permanent interest.

When they began to study the application of electricity to telegraphic purposes, one of the difficulties experienced was the tendency of the electric current to become weak as the length of the line and imperfect insulation increased. The currents were liable at comparatively short distances to become so weak as to be unable to produce intelligible signals. To remedy this defect the Brothers Siemens invented the relay—an electro-magnet so delicate that it will move with the weakest current. This simple apparatus is an adapiation to the telegraph wire of the principle by which a weak current can be converted into a strong one in the dynamo-machine. Although it simply looks like a small coil of wire, its power is such that it makes the current, however weak at first, increase in geometrical proportion. The Siemens polarised relay is the most perfect and powerful instrument of this description. By the use of five of them, each of which retransmits the original signals from a fresh battery, a message can be sent on the Indo-European Telegraph from London to Teheran, a distance of 3,800 miles, without any retransmission by hand.

Sir William Siemens, in company with his brother, Werner, and Herr Halske, established in 1858 the telegraphic works near London which are now known by the name of Siemens. A whole progeny of electrical apparatus invented by them are manufactured at these works, which sometimes employ 1,000 men. Some of the largest works in telegraphic engineering have also been produced there.

The construction of the Indo-European Telegraph—the first great undertaking of the kind—was undertaken by them. In May, 1867, the Messrs. Siemens obtained concessions for twenty-five years from the Prussian, Russian, and Persian governments, for an overland double line from England to India through Prussia, Southern Russia, and Persia. At that time very short messages cost 5*l.* ; they were sometimes a week in course of transmission, and were often unintelligible when transmitted. To provide better means of communication the Indo-European Telegraph Company was formed in 1868, with a Board of Directors that fairly represented the countries through which the lines would pass. To this Company the Messrs. Siemens transferred their concessions on condition that they should receive certain payments after the completed line paid a 12 per cent. dividend on the paid up capital. The Prussian government agreed to build the line at its own expense through North Germany, and the Messrs. Siemens contracted to build the whole of the line from Alexandrowo, near the Prussian frontier, to Teheran, for 400,000*l.*, and to maintain it for a further sum of 34,000*l.* a year. The length of this overland line was 2,700 miles; and they agreed to complete it in eighteen months. The tender was sent in on the 27th of April, 1868; it was accepted early in June; and the lines were completed, though not opened, on the 10th of December, 1869. The work of construction was nevertheless of no ordinary difficulty. The line goes from London *via* Lowestoft, Emden, Berlin, Warsaw, Jitomir, Odessa, Kertch, Sukhum, Tiflis, Tabriz, and Teheran, where it joins the Indian government lines to Bushire and Kurrachee. The materials for the line in Persia, consisting of 11,000 iron posts, 33,400 insulators, and 900 miles of wire of large section, were shipped to St. Petersburg, whence they were transported on the Neva and the Volga to Astrakhan, where they were again shipped across the Caspian for Lenkoran, Astara, and Resht, the northern ports of Persia. At these ports it was found difficult to get beasts of burden to distribute the materials in the interior of the country within the prescribed time. Nevertheless, all the difficulties encountered in an unsettled and uncommercial country were overcome, and the lines were completed in due time. But they were not opened till the 31st of January, 1871, and even then the wires were not in good working order. The chapter of accidents that caused these interruptions was well explained by the contractors in their reports to the directors. In these reports the Messrs. Siemens stated that "With the opening of the line winter weather of extraordinary severity set in in Persia and the south of Russia, commencing with sleet and heavy falls of snow, followed by intense cold, indicated by a fall of the thermometer of from 20° to 30° Reaumur below zero. The wires being weighted by a thick coating of sleet were drawn tight by the cold, and broke in hard places or at defective joints. Considering that the length of line wire exposed to these causes exceeds 5,000 miles, the number of these casualties has been extremely small, and would not have caused any sensible interruption of the service, notwithstanding the intense cold and the circumstance that the ground was deeply covered with snow, had not another disturbing cause presented itself. The interruptions have been entirely confined to Eastern Russia, whereas the Persian lines, though similarly circumstanced, have continued to be in good working condition, and this striking difference of results can only be attributed to the different construction of insulators used by us in Russia and in Persia. The Persian insulators are of a construction peculiar to ourselves, with cast-iron protecting caps inclosing an inverted porcelain bell, from the centre of which the line wire is suspended, whereas the insulator which we were *obliged* to use in Russia supports the line wire upon a bell of porcelain, mounted upon a metal stalk, held by an iron bracket. The latter description of insulators was insisted upon because they are the form more usually employed on European lines, and insulate well under ordinary circumstances, but they have two disadvantages in rough climates and uncivilised countries; namely, that they can be easily broken by stones, and further that snow rests upon the bell and bracket supporting the wire, and forms a conductive connection, or leak, between the line wire and the post—which disadvantages do not apply to our special insulator, where the insulating bell is protected under a strong iron cap, and where the line wire is suspended from the bell and presents no surface for the settlement of snow. These insulators are used also on the government lines in Persia, and have been adopted lately also upon the Turkish lines, with great advantage to the working of those lines; and considering the additional proof of their superiority, we hope we shall obtain permission to substitute them on your lines in Eastern Russia for at least one line wire—being willing to effect the change at our own expense—in the course of next summer, rather than run the risk of similar interruptions next winter.

" The maintenance of a considerable line of telegraphic communication during the first twelve months of its existence is always a task of some difficulty and disappointment. The wires composing the lines, however carefully prepared, will show hidden defects and occasional breakage, the posts will yield where the ground is treacherous or where the staying has been insufficient, and the insulators are liable to be wantonly destroyed by the mischievous persons of any community where the telegraph is a novelty; occasional interruptions of the service are the result, and are the more severely felt if no alternative lines are immediately available. These difficulties had to be expected, and were guarded against by the appointment of a considerable working staff of guards and superintendents of the line, whose duty it was not only to repair the line as soon as possible in cases of accident, but to prevent their recurrence by effecting local diversions or other improvements.

" The line having been opened, perhaps somewhat prematurely, during the worst part of the year 1870, the interruptions of the traffic were rather serious during the first month or two, but by pursuing the system above described we had succeeded in reducing their number and effect to such an extent that the through communication between London and Teheran was generally complete, and could be worked direct and instantaneously without intermediate repeating-stations, a result which has not been surpassed, we believe, in telegraphic practice; when, on the 7th of July, a calamity occurred which could not have been foreseen, namely, the destruction of both our land and cable lines in Georgia by a severe earthquake. Not only were the land lines thrown down and the wires torn at several points, which could easily have been set right, but the cable line through the Black Sea between Sotcha and D'juba, which had been successfully laid, and remained in excellent working condition up to that time, was suddenly torn in two places. A steamer, furnished with all the necessary appliances for repairing cables, was immediately despatched from Kertch, but in endeavouring to raise the cable it was found to be covered with earth at a point twenty miles distant from Sotcha, a result that could only be explained by a submarine landslip having taken place. It was evident that to repair the line more spare cable would have been required than was on board, and it would not have been possible to obtain a fresh supply of cable from England before the season would have been too far advanced to undertake a considerable repair operation on a boisterous and rocky coast. Moreover, it appeared, from inquiries on the coast, which had been but little known before our line was projected, that earthquakes of great severity had frequently been experienced, under which circumstances the submarine line would have remained in a hazardous condition after the repair would have been effected. On the other hand, the objection which had originally attached to this mountainous and deserted coast was in course of being removed by the construction of a coast-road, which the Russian government had in the meantime put in hand. After carefully weighing these circumstances, we came to the conclusion that the interests of the company would be best served by the construction of a substantial land line along the Caucasian coast. Applications were accordingly addressed to the Russian government to grant the necessary authority. The liberality with which the Russian government granted these requests made it possible for the Indo-European Telegraph Company to continue their telegraphic service after only a short interruption, and enabled us to push forward the new work, which was accomplished by the end of the year. On the 1st of January, 1871, messages again passed all the way from London to Teheran upon the company's own lines." Since then the line has worked well, and, although temporary interruptions are unavoidable, especially in Southern Russia, where the line is exposed to heavy storms, and to the accumulation in winter of masses of ice on the wires, they have generally been of short duration. From a commercial point of view, too, it is one of the most successful works of its kind.

The connection of the Messrs. Siemens with this undertaking came to a close in 1882. Writing in August of that year, when England's intense interest in the Egyptian campaign was attracting public attention to our means of communication with the East, Sir William Siemens said: "At the present time our communication with India, Australia, and the Cape depends, notwithstanding the nominal existence of a line through Turkey, on the Indo-European Telegraph. This line, referring now to the portion of the system connecting London and Teheran, with the origin and construction of which I have been intimately associated, has not been looked upon with much favour by many in this country, who at the time of its construction predicted its ultimate failure, and threw out broad hints to the effect that the telegraph posts might serve in certain regions to mark the tombs of the staff employed upon the work, while others took the objection that the line, if constructed, would be liable to frequent interruptions from political causes. I and those acting with me felt no misgivings on these points, because, before seeking to obtain concessions from Germany, Russia, and Persia for the construction of the line, we took the precaution of having its neutrality and independence from government interference guaranteed by an international convention between the two principal Powers concerned, which guarantee has been absolutely respected throughout the very trying times of the Franco-German and Russo-Turkish wars, as well as during the critical period of the subsequent peace negotiations at Constantinople, when the English despatches passed without hindrance over the Indo-European line *via* Odessa. At the present time the Indo-European telegraph is—not, indeed, for the first time—

practically the only means of communication between England and her Eastern possessions, nor does it prove itself insufficient or unreliable under these trying circumstances, land line though it be."

The Messrs. Siemens were also pioneers in submarine telegraphy. The first submarine telegraph cable covered with gutta-percha was laid by Dr. Werner Siemens in 1847 across the Rhine from Deutz to Cologne, a distance of half a mile. Previous attempts to effect insulation by resinous substances had failed. In the early days of submarine cables much difficulty was experienced in getting a suitable insulating covering which would effectually prevent the escape of the electric current through its entire length, as a single flaw was found to make a whole cable useless. Dr. Werner Siemens was the first to recommend and use gutta-percha or india-rubber, which was brought to England about that time; and its superiority soon became apparent, owing to its great tenacity and power of resisting heat. But the laborious operation by which india-rubber was applied as a covering to the wire made it expensive. In the first cables made in this way the india-rubber was cut into strips, which were wound spirally round the wires; this operation had to be repeated several times before the necessary degree of safety was attained; and though the overlapping edges were heated and soldered together, even then the covering was often imperfect. To obviate these difficulties Sir William Siemens invented a machine which combined the advantages of cheapness, quickness, and certainty of result. The machine was so constructed as to draw the india-rubber over the wire, and at the same time the newly-cut edges were united under pressure so as to combine firmly and form a secure covering. This invention was brought before the world in 1860.

Sir William subsequently designed the steamship *Faraday* specially for the work of laying submarine cables. This unique vessel was an improvement on all previous cable ships. For the first Atlantic cable that was laid the *Great Eastern*, the largest ship in the world, was remodelled in the interior, being fitted with three immense circular iron tanks which carried the cable; but as she was found costly to maintain and inconvenient to manage, experience showed the necessity of providing a vessel better adapted for the laying of long submarine cables.

The *Great Eastern* was used by the Telegraph Construction and Maintenance Company; the Hooper Telegraph Company built a vessel of their own for the same purpose, and called it the *Hooper;* and Sir William Siemens designed the most suitable vessel of all for his firm. It is 360 feet long, 52 feet wide, and 36 feet deep. It has a measured register of 5,000 tons, but is capable of carrying nearly 6,000 tons dead weight. It is built of iron, and is double bottomed, the spaces between the two bottoms consisting of a network of iron girders, the meshes of which are fitted to contain water ballast. In the interior of the ship are three enormous cable tanks constructed of plate iron, and so contrived as to form a series of double arches for supporting the sides of the ship; they are also united to one another and to the general fabric of the hull by five iron decks—an arrangement that makes the tanks a means of strengthening the ship instead of weakening it. Moreover, as the ship is lightened by discharging the cable and consuming the coals on board, water can be admitted into the cells between the double bottom to serve as ballast, thus keeping the vessel at a nearly uniform depth in the water. A complete system of valves, cocks, pipes, and other appliances, which are under the control of the engineers working in the engine-room, are used for filling and emptying any single compartment of the double bottom. The ship is made alike at both ends, and furnished with machinery capable of steering backward or forward with equal facility. The tanks are capable of storing 1,700 miles of cable $1\frac{1}{4}$ inch in diameter, and new and ingenious machinery is provided on deck for paying it out. The vessel is lighted by the electric light, whose perfect illumination enables the men on board to work day and night. All the heavy labour on board is performed by steam apparatus placed on various parts of the deck. There is also excellent cabin accommodation.

The two screws of this vessel are so constructed that she can turn in her own length when the engines, which are constructed with a view to great economy of fuel, are worked in opposite directions. On a voyage from Newcastle to London a cask was thrown overboard, and from it as a centre the vessel turned in her own length in 8 minutes 20 seconds, touching the cask three times during the operation. This manœuvring power is found to be of great importance in such a case as repairing a fault in a cable, as it enables the engineer to keep her head in position, and to place her just where necessary in defiance of side winds or currents.

This ship was called the *Faraday* in honour of the distinguished *savant* of that name. In speaking of the great service which Professor Faraday had rendered to electrical science, and the invariable kindness with which he had encouraged younger labourers in the same field, Sir William Siemens said the friendly encouragement which he himself had experienced from him would ever remain a most pleasing remembrance.

The *Faraday* was first used in laying the Direct United States Cable, which is above 3,000 miles in length. Nearly the whole of that cable, made of copper conductors and gutta-percha insulators, and a sheeting of steel wires covered with hemp, was laid in perfect condition in 1874, but in consequence of the stormy season setting in, its completion was postponed till 1875. In June of that year the *Faraday* resumed operations, and soon completed the work. But the discovery of a fault necessitated another return to England for a piece of cable to repair the damage. This delayed the opening of the cable till the 15th of September, on which day it was opened to the public for the transmission of messages. As regards construction, maintenance, and rate of transmission, the cable has been a great success. The same firm laid another transatlantic cable for the Compagnie Française du Télégraphe de Paris à New York with entire success in a surprisingly short space of time. The order for the cable was given by the French company in March 1879, and it was handed over to them in perfect working order in September of the same year. It transmits messages over a distance exceeding 3,000 miles. Even that feat has been eclipsed in the laying of subsequent cables.

The *Faraday* was one of the first vessels that used the electric light at sea; and as an illustration of the utility of that light in navigation, Sir William Siemens stated that in 1878 it saved a serious collision in the Atlantic. During a dense fog, the captain, standing on the bridge of the *Faraday*, saw by the electric light a dark mass moving before him, which he could not have seen with ordinary lights. At the same time the people on board the approaching vessel, which happened to be an emigrant ship, saw the electric light, although they would not have seen a common light. Both captains manœuvred their ships accordingly, and they just managed to escape each other—actually approaching within a yard—and thus prevented a collision, which would otherwise, in the decided opinion of the captain, have taken place. So impressed was Sir William with the value of the light on this occasion, that he wrote to the Board of Trade, whose regulations then forbade the use of the electric light on board ships, suggesting an interview between the captain of the *Faraday* and the Board of Trade authorities. The interview was immediately granted; but when the captain of the *Faraday* narrated the incident, he was met by the observation that he had committed an illegal act. His retort was, " But I saved the collision."

Sir William relates another incident that shows the " consistency " of the Board of Trade in another direction. He personally superintended all the mechanical arrangements in connection with the *Faraday ;* but with regard to the engines and boilers he depended entirely on the Board of Trade and Lloyd's rules, his instructions being simply to "make the boilers as safe and the engines as efficient as they can be made." The result, he says, was a success. The ship never failed in its arduous duties, often being for weeks together in winter in the Atlantic. In the course of a few years the Board of Trade rules were changed, and the boilers considered no longer sufficient. They were carefully inspected after each voyage, and reported to be in perfect condition by the surveyors both of the Board of Trade and of Lloyd's. Nevertheless, they reduced the pressure after each voyage five lbs. until the ship would soon have been dependent on her sails. Accordingly he had to put in steel boilers, at a cost of 10,000*l.*, while the old boilers were sold as secondhand, and went into ships not requiring to conform to the Board of Trade rules.

Though Sir William did not design the boilers first used in the *Faraday*, he afterwards invented a new form of boiler of surpassing lightness and strength. He described it to the Institution of Mechanical Engineers in 1878 as another proof of the superior properties of steel, as compared with iron, in resisting high pressure.

For some time previously the use of compressed air instead of steam in locomotives had been engaging the attention of engineers, as it was found that air locomotives could be successfully and economically used where steam could not. In collieries, for instance, the use of air locomotives was found to be cheaper than pony labour for carrying the trucks of coals from the workings to the bottom of the shafts. Colonel Beaumont was also experimenting with compressed air engines with a view to adapting them for tramways and underground railways, where steam and smoke are a nuisance. But, as Sir William stated, it was found difficult to construct a vessel capable of withstanding the great internal pressure necessary for such a purpose. In consequence of

the practical difficulties hitherto experienced in making such
vessels of boiler plate (iron), it was generally thought advisable to
limit the diameter of cylindrical vessels, and to resort to a multi-
tubular construction. But in these the seams of rivets and
many joints were sources of weakness; and such vessels necessarily
occupied much more room than a plain cylindrical vessel would
do. The use of cast-iron, too, in such vessels, in hydraulic
presses, and accumulators, required a degree of thickness that
rendered them extremely ponderous and costly; and it sometimes
happened that the fluid under pressure found its way through the
pores of the metal. In consequence of these difficulties, Colonel
Beaumont asked Sir William in 1877 to construct for him a vessel,
with a capacity of not less than a hundred cubic feet, capable of
resisting an internal pressure of at least 1,000 lbs. on the square
inch, and at the same time not exceeding two and a half tons in
weight. To meet these requirements Sir William used steel, made
at the Landore Steel Works, capable of resisting a tensile strain
of 45 tons per square inch and of extending from 8 to 10 per
cent. before breaking. Of that material he constructed a vessel
consisting of fourteen cylindrical rings of 40 inches internal
diameter, and 12 inches deep, rolled out of solid steel ingots, and
of two hemispherical ends beaten out of steel boiler plate. Two
rings of cast steel, each perforated with 20 holes, fitted over the
hemispherical ends, and through these holes were passed 20 bolts
of steel, capable of resisting 50 tons per square inch. The vessel
being built up of these parts, the bolts were gradually tightened
to a point just sufficient to resist the intended internal pressure.
It was then filled with water, and the pressure of a hydraulic
accumulator loaded to 1,000 lbs. per square inch was applied.
With this test it showed no sign of leakage. The internal pressure
was therefore raised to 1,300 lbs. per square inch, at which point
nearly all the joints began to weep, showing that the bolts were
beginning to elongate. Upon drawing up each nut another eighth

of a turn, the vessel was found perfectly tight at 1,300 lbs. per square inch, but it began to weep again when the pressure was raised to 1,400 lbs. "Considering," said the inventor, "that the intended working pressure of this vessel was only 1,000 lbs. per square inch, it was thought unnecessary to draw the bolts any tighter, although, according to calculation, the rings as well as the bolts were capable of resisting with safety above 2,000 lbs. per square inch. The great length of the bolts insured a sufficiently elastic range of action for this purpose, and being made of steel containing one half per cent. of carbon, they would retain their elasticity for an indefinite length of time. An hydraulic press constructed on this principle should not weigh more than one-fourth of the weight of a press of the ordinary construction. A boiler of this construction possesses, in common with the air vessel just described, the advantage of leaking, through the yielding of the elastic bolts, long before there is the least danger of explosion. It possesses, moreover, the additional advantage that it can be carried in pieces to be put together *in situ*, thus facilitating carriage and avoiding the necessity of providing hatchways of extraordinary dimensions for putting the boilers on board."

Sir William Siemens has been one of the most versatile inventors in England. Though he has taken out more than one hundred patents of his own, all his inventions have not been patented. His first patent was taken out in 1845, and rarely has a year elapsed since then without one invention or more being recorded by him in the patent office. In addition to the patents that are exclusively his own, there are a good many in the joint names of the Brothers Siemens, Sir William having always shown the most scrupulous care in giving his brothers full credit for their share in inventions of which they are joint patentees. There are forty or fifty of that description.

Some of his inventions which have never been patented are

recorded in the Proceedings of learned societies instead of in the Patent Office. For example, in 1866 he read a paper before the Royal Society on Uniform Rotation, in which he described a new kind of governor, called the gyrometric governor. He stated that some months previously there occurred to him an idea which, while it furnished the elements of a very general and com- plete solution of the problem of uniform rotation, appeared to possess also a separate scientific interest. An open cylindrical glass vessel or tumbler containing some liquid being made to rotate upon its vertical axis, he observed that the liquid rose from the centre towards the sides to a height depending on the angular velocity of the diameter of the vessel. As soon as the velocity reached a certain limit the liquid commenced to overflow the upper edge of the vessel, being thrown from it in the form of a liquid sheet in a tangential direction. If the velocity remained constant, the overflow of the liquid ceased, although it continued to touch the extreme edge or brim. When the velocity of the vessel was diminished, the liquid was observed to sink, and to rise again to its former position when the rotation was raised to its previous limit of angular velocity. This velocity was the result of the balance of two forces acting on the liquid particles; namely, gravity and centrifugal force. He applied this principle to the regulation of steam-engines and other machines where the nearest approach to uniform rotation was desirable. The rotating vessel constructed for this purpose consisted of a cup open at both top and bottom, but widest at the top. The narrow bottom was placed in another vessel containing water, which it just touched while by mechanical appliances the cup itself was made to revolve at a velocity proportionate to the strength of the motive power employed. He found that rotation being thus imparted to the cup, the liquid rose in it by centrifugal force, while additional liquid entered from without and maintained the apex of the liquid curve. Experiments made with this apparatus showed that the driving power might be varied between the widest limits without producing any sensible variation of speed. The final adjustment of the instrument to the normal velocity required was, moreover, easily effected by raising or lowering the cup while it was running, for which purpose mechanical appliances were provided. To illustrate the application of this principle, he constructed a clock which was driven by electro-magnetism, whose power was regulated by the cup, while a train of reducing wheels communicated the motion of the cup to the face of the clock, which recorded the hours and minutes in the usual manner. In its application to steam-engines, the most striking feature of this governor, said its inventor, was the rapidity with which the re-adjustment between the power and the load of the engine was effected. He proved by experiment that two-thirds of the total load upon an engine could be suddenly thrown off without producing any visible change in its rotation. The paper recording these experiments was ordered to be printed in the *Philosophical Transactions.*

In 1871 he contributed a paper to the Royal Society "On Electrical Resistance," which was made the Bakerian lecture for that year. It explained a method of measuring variation of temperature by variation of electrical resistance ; and described two new instruments—the electrical resistance thermometer and pyrometer, in connection with the differential voltameter—which he invented, and which are now recognised as ingenious and useful aids in thermometry and metallurgy. These instruments can measure temperature without any break from the lowest possible degree of cold to a temperature approaching that of the fusion of platinum. Many eminent men of science have endeavoured to invent such an instrument during the last 150 years ; but Sir William Siemens, who studied the question more or less for ten years, was the first to construct a reliable pyrometer of unlimited range and universal application. Its first application was the means of saving an important telegraph cable from destruction through spontaneous generation of heat. It has been used for recording the temperature at elevated points and at points below the earth's surface. It has been successfully used for determining the internal temperature of the blast furnace, and recording the same in the ironmaster's office, sometimes situated at a distance from the blast furnace. It has also been used for ascertaining the temperature of the bottom of the ocean.

Again, while he has been unceasing in his efforts to impress upon the public the scope there is for economy of fuel, he in 1879 constructed a form of fire-grate that brought the means of effecting this economy, more or less, within the power of every householder; but in order that it might be used without restraint and at the least expense, he did not make it the subject of a patent.

In later years Sir William Siemens again came before the world with some further results of his life-long study of questions relating to combustion and the utilisation of different forms of energy. Some of the conclusions he arrived at on these questions have excited the wonder and criticism of the greatest scientific men in all parts of the world. His own expositions of the subject are remarkable for simplicity and originality. The following extract from a lecture on "Fuel," which he delivered at Bradford in 1873, under the auspices of the British Association, is an admirable introduction to his views on this subject. He then stated that "fuel in the ordinary acceptation of the term is carbonaceous matter, which may be in the solid, the liquid, or the gaseous condition, and which, in combining with oxygen, gives rise to the phenomena of heat. Commonly speaking, this development of heat is accompanied by flame, because the substance produced in combustion is gaseous. In burning coal, for instance, in a fire-grate, the oxygen of the atmosphere enters into combination with the solid carbon of the coal and produces carbonic acid, a gas which enters the atmosphere, of which it forms a necessary constituent, since without it the growth of trees and other plants would be impossible. But combustion is not necessarily accompanied by flame, or even by a display of intense heat. The metal magnesium burns with a great display of light and heat, but without flame, because the product of combustion is not a gas but a solid, viz., oxide of magnesia. Again, metallic iron, if in a finely divided state, ignites when exposed to the atmosphere, giving rise to the phenomena of heat and light without flame, because the result of combustion is iron oxide or rust ; but the same iron, if presented to the atmosphere—more especially to a damp atmosphere—in a solid condition, does not ignite, but is nevertheless gradually converted into metallic oxide or rust, as before. Here, then, we have combustion without the phenomena either of flame or light; but by careful experiment we should find that heat is nevertheless produced, and that the amount of heat so produced precisely equals that obtained more rapidly in exposing spongy iron to the action of oxygen. Only in the latter case the heat is developed by slow degrees, and is dispersed as soon as produced, whereas in the former the rate of production exceeds the rate of dispersion, and heat therefore accumulates to the extent of raising the mass to redness. It is evident from these experiments that we have to widen our conception, and call fuel any substance which is capable of entering into combination with another substance, and in so doing gives rise to the phenomenon of heat.

" In looking at the solid crust of the earth we find it to be composed for the most part of siliceous, calcareous, and magneseous rock. The former, silica, consists of the metal silicon combined with oxygen, and is therefore not fuel, but rather a burnt substance which has parted with its heat of combustion ages ago. The second, limestone, is carbonate of lime, or the combination of two substances, viz., oxide of calcium and carbonic acid, both of which are essentially products of combustion, the one of the metal calcium, and the other of carbon. The third is the substance magnesium, which, combined with lime, constitutes dolomite rock, of which the Alps are mainly composed. All the commoner metals, such as iron, zinc, tin, alumina, sodium, &c., we find in nature in an oxidised or burned condition; and the only metallic substances that have resisted the intense oxidising action that must have prevailed at one period of the earth's creation are the so-called precious metals, gold, platinum, iridium, and, to some extent also, silver and copper. Excepting these, coal alone presents itself as carbon and hydrogen in an oxidised condition. But what about the oceans of water which have occasionally been cited as representing a vast store of heat-producing power ready for use when coal shall be exhausted? Not many months ago statements to this effect could be seen in some of our leading papers. Nothing, however, could be more fallacious. When hydrogen burns doubtless a great development of heat ensues, but water is already the result of this combustion (which took place upon the globe before the ocean was formed), and the separation of these two substances would take precisely the same amount of heat as was originally produced in their combustion. It will thus be seen that both the solid and fluid constituents of our earth, with the exception of coal, of naphtha (which is a mere modification of coal), and the precious metals, are products of combustion, and therefore the very reverse of fuel. Our earth may indeed be looked upon as 'a ball of cinder, rolling eternally through space,' but happily in company with another celestial body—the sun—whose glorious beams are the physical cause of everything that moves and lives, or that has the power within itself of imparting life, heat, or motion. Its invigorating influence is made perceptible to our senses in the form of heat."

After explaining that combustion ceases, according to Sainte-Claire Deville, at 4,500° Fahr., which has been called the point of dissociation—because at that point hydrogen might be mixed with oxygen and yet the two would not combine or produce combustion—he went on to say :

" All available energy upon the earth, excepting the tidal wave, is derived from the sun, and the amount of heat radiated year by year upon our earth could be measured by the evaporation of a layer of water fourteen feet thick spread over the entire surface, which again would be represented by the combustion of a layer of coal one foot in thickness covering our entire globe. It must, however, be taken into account that three-fourths of this heat are intercepted by our atmosphere, and only one-fourth reaches the earth itself. The amount of heat radiated away from the sun would be represented by the annual combustion of a thickness of coal seventeen miles thick covering its entire surface ; and it has been a source of wonder with natural philosophers how so prodigious an amount of heat could be given off year after year without any appreciable diminution of the sun's heat having become observable. Recent researches with the spectroscope, chiefly by Mr. Norman Lockyer, have thrown much light upon this question. It is now clearly made out that the sun consists near the surface, if not throughout its mass, of gaseous elementary bodies, and in a great measure of hydrogen gas, which cannot combine with the oxygen present, owing to great elevation of temperature (due to the original great compression) which has been estimated at from 20,000° to 22,000° Fahr. This chemically inert and comparatively dark mass of the sun is surrounded by the photosphere, where the gaseous constituents of the sun rush into combustion, owing to reduction of temperature in consequence of their expansion and of radiation of heat into space. This photosphere is surrounded in its turn by the chromosphere, consisting of the products of combustion, which, after being cooled down through loss of heat by radiation, sink back, owing to their acquired density, towards the centre of the sun, where they become again intensely heated through compression, and are 'dissociated' or split up again into their elements at the expense of internal solar heat. Great convulsions are thus continually produced upon the solar surface, resulting frequently in explosive actions of extraordinary magnitude, when masses of living fire are projected a thousand miles or more upward, giving rise to the phenomena of sun-spots and of the corona, which is visible during the total eclipses of the sun. The sun may therefore be looked upon in the light of a gigantic gas furnace in which the same materials of combustion are used over and over again."

Continuing his study of this stupendous problem, he brought his matured thoughts on the subject before the Royal Society in March, 1882. He then stated that the amount of heat radiated from the sun had been approximately computed at 18,000,000 of heat units for every square foot of his surface per hour, or as equal to the heat that would be produced by the perfect combustion every thirty-six hours of a mass of coal as great as that of our earth. If the sun were surrounded by a solid sphere with a radius equal to the mean distance of the sun from the earth (93,000,000 miles), the whole of this prodigious amount of heat would be intercepted; but considering that the earth's apparent diameter as seen from the sun is only seventeen seconds, the earth can intercept only the 2,250-millionth part. . . . The sun completes one revolution on his axis in twenty-five days, and his diameter being taken at 882,000 miles, it follows that the tangential velocity amounts to $1\frac{1}{4}$ mile per second, or nearly $4\frac{1}{2}$ times that of our earth. The high rotative velocity of the sun must cause an equatorial rise of the solar atmosphere. If solar rotation takes place within a medium of unbounded extension, the sun would act mechanically upon the floating matter surrounding him in the manner of a fan, drawing it towards himself upon the polar surfaces, and projecting it outward in a continuous disk-like stream. By this fan-like action hydrogen, hydrocarbons, and oxygen are supposed to be drawn in enormous quantities toward the polar surfaces of the sun; during their gradual approach they will pass from their condition of extreme attenuation and extreme cold to that of compression, accompanied with rise of temperature, until on approaching the photosphere they burst into flame, giving rise to a great development of heat, and a temperature commensurate with their point of dissociation at the photospheric density. The result of their combustion will be aqueous vapour and carbonic anhydride or oxide, according to the sufficiency or insufficiency of oxygen present to complete the combustion; and these products of combustion, in yielding to the influence of propulsive force, will flow toward the solar equator, and be thence projected into space. . . . By means of the fan-like action resulting from the rotation of the sun, the vapours dissociated in space would be drawn toward the polar surfaces of the sun, be heated by increase in density, and would burst into a flame at a point where both their density and temperature had reached the necessary elevation to induce combustion, each complete cycle taking years or centuries to be accomplished. The resulting aqueous vapour, carbonic anhydride and carbonic oxide, would be drawn towards the equatorial regions, and be then again projected into space by centrifugal force.

In support of these views Sir William Siemens made numerous experiments which showed the application of these physical principles on a small scale. Experiment and reflection led him, he says, to look upon the sun in the light of a vast piece of apparatus, worked upon principles that could be observed and appreciated at their real value in terrestrial practice. Many scientific authorities disputed this theory of solar action, but none of the arguments used against it seemed powerful enough to shake his faith in it. On the contrary, evidences in favour of it continued to accumulate. Referring to this subject in his presidential address at the British Association in August, 1882, he stated that, "armed with greatly improved apparatus, the physical astronomer has been able to reap a rich harvest of scientific information during the short periods of the last two solar eclipses—that of 1879, visible in America, and that of May, 1881, visible in Egypt by Lockyer, Schuster, and Continental observers of high standing. The result of this last eclipse expedition has been summed up as follows :—" Different temperature levels have been discovered in the solar atmosphere ; the constitution of the corona has now the possibility of being determined, and it is proved to shine with its own light. A suspicion has been aroused once more as to the existence of a lunar atmosphere, and the position of an important line has been discovered. Hydrocarbons do not exist close to the sun, but may in space between us and it."

"To me personally these reported results possess peculiar interest, for in March last I ventured to bring before the Royal Society a speculation regarding the conservation of solar energy, which was based upon the three following postulates, viz. :—

1. That aqueous vapour and carbon compounds are present in stellar or interplanetary space.

2. That these gaseous compounds are capable of being dissociated by radiant solar energy while in a state of extreme attenuation.

3. That the effect of solar rotation is to draw in dissociated vapours upon the polar surfaces, and to eject them after combustion has taken place back into space equatorially.

"It is therefore a matter of peculiar gratification to me that the results of observation here recorded give considerable support to that speculation. The luminous equatorial extensions of the sun which the American observations revealed in such a striking manner (with which I was not acquainted when writing my paper)

P

were absent in Egypt; but the outflowing equatorial streams I suppose to exist could only be rendered visible by reflected sunlight, when mixed with dust produced by exceptional solar disturbances or by electric discharge; and the occasional appearance of such luminous extensions would serve only to disprove the hypothesis entertained by some, that they are divided planetary matter, in which case their appearance should be permanent. Stellar space filled with such matter as hydrocarbon and aqueous vapour would establish a material continuity between the sun and his planets, and between the innumerable solar systems of which the universe is composed. If chemical action and reaction can further be admitted, we may be able to trace certain conditions of thermal dependence and maintenance, in which we may recognise principles of high perfection, applicable also to comparatively humble purposes of human life."

In 1877 Sir William Siemens, in his inaugural address as president of the Iron and Steel Institute, again called attention to the consumption and waste of fuel; and to show the magnitude of power which is now for the most part lost, but which may be sooner or later called into requisition, he took the Falls of Niagara as a familiar example. He said : " The amount of water passing over this fall has been estimated at 100,000,000 of tons per hour, and its perpendicular descent may be taken at 150 ft., without counting the rapids, which represent a further fall of 150 ft., making a total of 300 ft. between lake and lake. But the force represented by the principal fall alone amounts to 16,800,000 horse-power, an amount which, if it had to be produced by steam, would necessitate an expenditure of not less than 266,000,000 tons of coal per annum, taking the consumption of coal at four pounds per horse-power per hour. In other words, all the coal raised throughout the world would barely suffice to produce the amount of power that continually runs to waste at this one great fall. It would not be difficult, indeed, to realise a large proportion of the power so wasted, by means of turbines and water-wheels erected on the shores of the deep river below the falls, supplying them from races cut along the edges. But it would be impossible to utilise the power on the spot, the district being devoid of mineral wealth or other natural inducements for the establishment of factories. In order to render available the force of falling water at this and hundreds of other places similarly situated, we must devise a practicable means of transporting the power. Sir William Armstrong has taught us how to carry and utilise water-power at a distance, if conveyed through high pressure mains. Time will probably reveal to us effectual means of carrying power to great distances, but I cannot refrain from alluding to one which is, in my opinion, worthy of consideration, namely, the electrical conductor. Suppose water-power to be employed to give motion to a dynamo-electrical machine, a very powerful electrical current will be the result, which may be carried to a great distance, through a large magnetic conductor, and then be made to impart motion to electro-magnetic engines, to ignite the carbon points of electric lamps, and to effect the separation of metals from their combinations. A copper rod three inches in diameter would be capable of transmitting 1,000 horse-power a distance of, say, thirty miles—an amount sufficient to supply one-quarter of a million candle-power, which would suffice to illuminate a moderately sized town."

This passage has been often quoted in the current literature of the day. It has been much used and much abused. Returning to the subject in 1881, Sir William Siemens said : " When, only five years ago, in addressing the Iron and Steel Institute, I ventured upon the assertion that the time was not distant when the great natural sources of power, such as waterfalls, would be transferred to considerable distances by means of stout electric conductors, to be there utilised for providing towns with light and motive power, I elicited an incredulous smile even from some of those most conversant with the laws of electricity. I could now point to at least three instances in this country where power is practically transmitted to a distance by means of electricity, to be utilised for pumping water, for lighting, for working machinery, and for the transmission of locomotive power."

The application of the same idea formed the chief subject of Sir William Thomson's presidential address to the Mathematical and Physical Science Section of the British Association in 1881. He said : " The splendid suggestion made about five years ago by Sir William Siemens, in his presidential address to the Iron and Steel Institute, that the power of Niagara might be utilised by transmitting it electrically to great distances, has given quite a fresh departure for design in respect to economy of rain-power. With the idea of bringing the energy of Niagara usefully to Montreal, Boston, New York, and Philadelphia, I calculated the formula for a distance of 300 British statute miles (which is greater than the distance of any of those four cities from Niagara, and is the radius of a circle covering a large and very important part of the United States and British North America), and I found almost to my surprise that, even with so great a distance to be provided for, the conditions are thoroughly practicable with good economy, all aspects of the case carefully considered." Again, in 1883, Sir William Siemens, speaking before the Institution of Civil Engineers, said it would be interesting to test his early calculation by recent experience. Mr. Marcel Deprez had lately succeeded in transmitting as much as three horse-power to a distance of twenty-five miles through a pair of ordinary telegraph wires of 4 mm. diameter. The results so obtained had been carefully noted by Mr. Tresca, and had been communicated to the French Academy of Sciences. Taking the relative conductivity of iron wire employed by Deprez, and the three-inch rod proposed by Sir William, the amount of power that could be transmitted through the latter would be about 4,000 horse-power.

In the land of his adoption his labours and genius have not

been allowed to go unrewarded or unhonoured. He has received many medals and honours from learned societies. The Society of Arts presented him with its gold medal for his regenerative condenser in 1850, and the Institute of Civil Engineers awarded him the Telford medal in 1852 for his paper " On the Conversion of Heat into Mechanical Effect." At the London Exhibition in 1862, and at the Universal Exhibition in Paris in 1867, he received prize medals for his regenerative gas-furnace and steel process. In 1874 he received the Royal Albert Medal in recognition of his scientific researches and inventions in connection with heat and metallurgy; and with the same object the Bessemer gold medal of the Iron and Steel Institute was presented to him in 1875. Oxford University conferred the degree of D.C.L. upon him in 1869; Glasgow University likewise bestowed the degree of LL.D.; and he was made a F.R.S. in 1862. He was also a prominent member of many learned societies. He has been president of the Institution of Mechanical Engineers, of the Society of Telegraph Engineers, of the Iron and Steel Institute, of the Society of Arts, and of the British Association for the Advancement of Science. He was a member of the Council of the Institution of Civil Engineers, of the Iron and Steel Institute, of the British Association, of the Royal Institution, and the Royal Society. To all of these societies he contributed valuable papers on scientific subjects.

It is sad to have to add that a career so distinguished and useful came to an abrupt close. While walking in Piccadilly on the 5th November, 1883, he accidentally fell on the pavement, and from the injuries then received he never recovered. After a fortnight's illness he died from rupture of the nerves of the heart, and his death was mourned as a national loss.

SIR JOSEPH WHITWORTH.

CHAPTER VIII.

" As we perceive the grass to have grown, but did not see it growing, and
as we perceive the shadow to have moved along the dial, but did not see it
moving ; so the advances we make in knowledge, as they consist of such
minute steps, can only be measured by the distance."—ADDISON.

IF it be true, as Carlyle has said, that "we are to bethink us
that the epic verily is not Arms and the Man, but Tools and the
Man—an infinitely wider kind of epic," he has himself supplied
the key-note of the epic of the future. " Man," he says, "is a
tool-using animal. Weak in himself, and of small stature, he
stands on a basis, at most for the flattest soled of some half square
foot, insecurely enough ; has to straddle out his legs, lest the
very wind supplant him. Feeblest of bipeds ! Three quintals are
a crushing load for him ; the steer of the meadow tosses him aloft,
like a waste rag. Nevertheless he can use tools, can devise tools :
with these the granite mountain melts into light dust before him ;
he kneads glowing iron, as if it were soft paste ; seas are his smooth
highway, winds and fire his unwearying steeds. Nowhere can you
find him without tools; without tools he is nothing, with tools
he is all." In the epic of Tools, surely Sir Joseph Whitworth will
be one of the foremost heroes. The first inventor who took out
a patent for making machine tools, no man has done more to
promote efficiency and facility in the use of workshop appliances.
" The most celebrated mechanician of this country," he has

invented a whole set of machine tools that have revolutionised the labours of our workshops ; he has invented means for improving the metal of which tools are usually made ; and he has provided with princely liberality for the improvement of that technical education upon which depends the success of future generations in the skilful use of tools.

Carlyle, who was capable of going into raptures over tools, has given us a definition of genius which is generally considered the reverse of poetic, and is sometimes regarded as anything but complimentary. Is it possible that after all the incense that for ages has been offered at the shrine of genius, the " celestial fire " which princes worshipped and poets prayed for has in the age of tools been transformed into " an infinite capacity for taking pains ? " Surely not. Marvellous things still continue to be said of genius even in this age of tools. For example, it is recorded of William Rowan Hamilton, who died in 1867, that when only three years of age he was a superior reader of English and well advanced in arithmetic ; at four he had a good knowledge of geography ; at five he was able to read and translate Latin, Greek, and Hebrew, and fond of reciting Dryden, Collins, Milton, and Homer ; at eight he had learned Italian and French ; at nine he studied Arabic and Sanscrit ; at eleven he compiled a Syriac grammar ; and at thirteen he could write letters in Persian. This youth afterward became the Royal Astronomer for Ireland, and gave to the world the principle of quaternions.

Not less surprising were the early intellectual powers displayed by Henry Smith, who died in 1883. At the age of two years he was able to read ; and at the age of four he was found teaching himself Greek from an old-fashioned grammar full of antique contractions in the characters. His mother carried on his education till he was eleven, and after that he was under a tutor, who has stated that within nine months afterwards, the boy read all Thucydides, Sophocles, and Sallust, twelve books of Tacitus, the

216 THE CREATORS OF THE AGE OF STEEL.

greater part of Horace, Juvenal, Persius, and several plays of Æschylus and Euripides; he also got up six books of Euclid and algebra to simple equations; he read a considerable quantity of Hebrew; and among other things, he learnt all the Odes of Horace by heart. Such was the intellectual childhood of the man who afterward became the greatest mathematician at Cambridge University.

Again, of James Clerk Maxwell, who died in the winter of 1879, it is said that before he was three years old he was busily engaged in investigating the mysteries of the bell-wires which ran through his father's house, and on a starry winter evening he was wrapped in a plaid and carried to the hall-door by his father, who there gave him his first lessons in astronomy. In his school-boy days he became absorbed in experiments on the compression of solids and the composition of light. In his fourteenth year he wrote a paper on oval curves, which was pronounced on the highest scientific authority to be worth reading before the Royal Society of Edinburgh, on account of the simplicity and elegance of its method as well as its originality. He grasped almost intuitively results which cost others infinite pains. A fellow-student at Cambridge who spent midnight and early morning hours in preparation for Mr. Hopkins, the mathematical tutor, says that half an hour or so before the time, Maxwell would rise and cheerfully say, "Well, I must go to old Hop's problems." Yet his work was always so well done that Mr. Hopkins said he never knew Maxwell to make a mistake. At the age of nineteen, to the astonishment of members who did not know the slender stripling, he disputed some point in the colour theory with Sir David Brewster at the British Association. At the age of 48 he died; yet at that early age he had made discoveries in physical science that "enriched the inheritance left by Newton, consolidated the work of Faraday, and impelled the mind of Cambridge University to a fresh course of real investigation."

Would it not be popularly regarded as the language of depre-
ciation to speak of such amazing powers as merely an infinite
capacity for taking pains? Yet, however disenchanting the
admiccion may be, the "infinite capacity for taking pains" will
be found to have played a most important part in the epic of
"Tools and the Man." Even the transcendent genius "who
first carried the line and plummet to the outskirts of creation"
was not above "taking pains." "Accurate and minute measurement," says Sir William Thomson, "seems to the non-scientific
imagination a less lofty and dignified work than looking for
something new. But nearly all the grandest discoveries of
science have been but the rewards of accurate measurement
and patient long continued labour in the minute sifting of numerical
results. The popular idea of Newton's grandest discovery is that
the theory of gravitation flashed into his mind, and so the discovery
was made. It was by a long train of mathematical calculation,
founded on results accumulated through the prodigious toil of
practical astronomers, that Newton first demonstrated the forces
urging the planets towards the sun, determined the magnitude of
those forces, and discovered that a force following the same law
of variation with distance urges the moon towards the earth.
Then first, we may suppose, came to him the idea of the
universality of gravitation; but when he attempted to compare
the magnitude of the force on the moon with the magnitude of
the force of gravitation of a heavy body of equal mass at the earth's
surface, he did not find the agreement which the law he was
discovering required. Not for years after would he publish his
discovery as made. It is recounted that, being present at a meeting
of the Royal Society, he heard a paper read describing geodesic
measurement by Picard, which led to a serious correction of the
previously accepted estimate of the earth's radius. This was what
Newton required. He went home with the result, and commenced
his calculations, but felt so much agitated that he handed over the

arithmetical work to a friend; then (and not when, sitting in a garden, he saw an apple fall,) did he ascertain that gravitation keeps the moon in her orbit."

It thus appears that the greatest achievements in the realm of physical truth may become common-place when stripped of the halo of romance in which tradition has enveloped them; and thus it is that labourers in the field of science or mechanics may, as Goldsmith said of literary men, lead uneventful lives. "Taking pains," even infinite pains, is rarely a fascinating operation; indeed, it is often the reverse. Hence the results obtained by men of this stamp are generally of more value than a detailed record of the labours that preceded them, even if, as is often the case with them, it is labour refreshed with hope and crowned with success. In this category may be ranked the life's work of Sir Joseph Whitworth. One of his most discriminating admirers has stated that for his eminence in his profession Sir Joseph is indebted to a natural aptitude for its cultivation, but mainly " to an exhaustive knowledge of its principles and processes, acquired by unconquerable perseverance. He does not belong to the ordinary type of inventors,— quick, versatile, and ingenious, acting from impulse or apparent inspiration; his productions, on the contrary, are the results of slow and deliberate thought, bringing former observation to bear upon tentative experiment, and accepting nothing as established till it has undergone proof. Proceeding with a logical severity to rest further operations only on ascertained facts, his process is so strictly inductive that he might justly be designated the Bacon of mechanics."

Joseph Whitworth was born at Stockport, in Cheshire, on the 21st of December, 1803. He was educated by his father till he was twelve years old, and he was then sent to a private academy at Idle, near Leeds, where he remained eighteen months. He then left school, and entered the manufactory. He was placed under the care of an uncle, who had a cotton mill in Derbyshire, wherein, while still a boy of fourteen years, he began to acquire a knowledge of machinery. At the age of eighteen he left the mill and entered the workshop. In the capacity of a mechanic, he worked for four years in Manchester, where he gave every attention to the making of machinery, which may be said to have then become the object of his life. At the age of twenty-one he removed to London, where he worked for eight years, chiefly with Maudslay & Clements, the former of whom was himself an inventor of tools, and the latter was associated with Babbage in the construction of his calculating machine. By this laborious training, and in this school of inventors, he acquired that practical knowledge and mechanical skill, which, aided by his inventive faculties, afterwards yielded such abundant results. In those days the working mechanic was almost unaided by machinery; the planing machine was then unknown; the chisel and hammer, the file, the primitive lathe, and a simple screw-cutting machine formed the whole repertory of engineers' tools. The measure then in use was a two-foot rule divided into eighths of an inch; and though the skilled workmen ventured to go to 16ths, 32nds, bare 32nds, and full 32nds, the general results were of a rude and primitive kind, and work of a high character could only be produced by superior skill on the part of the workmen. It was in these circumstances that young Whitworth early saw the great need there was for improved tools and machinery. In the cotton spinning and weaving factories around Manchester there was a growing demand for mechanical work of a higher character to cheapen and improve the complicated machinery which was then in use, but which it was difficult to make. His early connection with the cotton trade impressed him with the great advantages to be gained by improvements in the means of producing more perfect and exact machinery. Among the many skilled workmen employed in Maudslay & Clements' workshops, he was esteemed the best; but not content with personal distinction as a mechanic, he set himself to the production of machine tools —machines made for the purpose of producing other machinery. To this work he devoted his years from attaining manhood to middle life. It has been stated as a general law in scientific thought, that the best and most original ideas have always been conceived before the age of thirty. However generally this may be the case, it is the fact that before Sir Joseph Whitworth was thirty years of age he succeeded in producing plane surfaces with a degree of precision which was then unknown, and this formed the groundwork of nearly all his other machines. This achievement was perfected and described by him in 1840. The old method of producing true planes was that of grinding the surfaces of plates alternately with emery powder and water, which was as imperfect as it was laborious. Grinding, according to his own account,[1] was objectionable because it was unreliable. If one plane was true and the other not, grinding might give part of the error to both, instead of the true plane being imparted to each. Moreover, grinding powder collected in greater quantity about the edges of the metal than in the interior parts, producing untrue planes. The practice of grinding altogether impeded the progress of improvement. A true surface, instead of being in common use, was almost unknown. The want of it in various departments of the arts and manufactures was sensibly felt. The valves of steam-engines, the tables of printing-presses, stereotype plates, surface-plates, slides of all kinds required a high degree of truth, much superior to what they generally possessed. In the preparation of a surface-plate where there was already a model, generally the method of insuring the true nature of the new one was by spreading thin colouring matter over it, and then rubbing it with the true plate. The colour formed a thin film over the brightness of the plate, and where it was not true, the colour did not make a full impression, the higher points on the rising surface being coloured

---

[1] Paper read at the British Association meeting in Glasgow, 1840.

over, while the other parts were left in the shade. This dappled appearance showed the precise conditions of the new surface in every part, and enabled the mechanic to make it correspond with the original.

Such was the primitive process which was superseded by a method of mechanical precision. After repeated efforts, involving much labour and ingenuity, to produce true planes by means of the steel straight-edge and scraper, he succeeded, in 1830, in " originating " the first true planes ever made. The work of producing copies of them sufficiently accurate for workshop purposes then became comparatively easy. One of the best and most successful machines for this work is his edge planing machine, which can be made of almost any size. Sir Joseph had one made in his own works capable of taking a single cut forty feet in length. The bed of this machine is fifty feet long, its grooves being considered the longest true planes that have ever been made. So exactly can surface plates be made by his apparatus that if one of them be placed upon another, when clean and dry, the upper one will appear to float upon the lower one without being actually in contact with it, the weight of the upper plate being insufficient to expel, except by slow degrees, the thin film of air between their surfaces. But if the air be expelled, the plates will adhere together, so that by lifting the upper one the lower will be lifted along with it, as if they formed one plate.

To test this property of true planes before a meeting at the Royal Institution, Professor Tyndall exhibited two exceedingly accurate hexagonal planes which remained adherent in the best vacuum obtainable by a good air pump. The atmosphere was reduced until its total pressure on the surface of the hexagon amounted to only half a pound. The lower plate weighed 3 lbs., and to it was attached a mass of lead weighing 12 lbs. Though the pull of gravity was here thirty times the pressure of the atmosphere, the weight was supported. Indeed, it was obvious, says the learned Professor, when an attempt was made to pull the plates asunder, that had a weight of 100 lbs. instead of 12 lbs. been attached to the lower hexagon, it also would have been sustained by the powerful attraction of the two surfaces. "To show the probable character of the contact between the planes, two very perfect surfaces of glass were squeezed together with sliding pressure. They clung, apparently, as firmly as the Whitworth planes. Throwing, by reflection from the glass plates, a strong beam of light upon a white screen, the colours of " thin plates " were vividly revealed. Clasping the plates of glass by callipers and squeezing them, the colours passed through various changes. When monochromatic light was employed, the successions of light and darkness were numerous and varied, producing patterns of great beauty. All this proved that, though in such close mechanical contact, the plates were by no means in optical contact, being separated by distances capable of embracing several wave-lengths of the monochromatic light."

The economy of this invention is not less than its mechanical precision; the previous cost of planing surfaces, done by hand, was 12s. a foot; and now with this machine it is only a penny a foot. As showing the different conditions of the trade, however, it should be added, on the authority of the inventor himself, that when the cost of labour for making a true plane by chipping and filing was 12s. per foot, the capital required for tools for one workman was only a few shillings, but now, when the labour is lowered to 1d. per foot, the capital required for planing machines often amounts to 500l.

It was while working as a jouneyman mechanic that Sir Joseph Whitworth conceived the idea of making true planes; and having partly accomplished this task he returned from London to Manchester in 1833, and began business on his own account. Over his door was written " Joseph Whitworth, tool-maker, from

London," and this was the beginning of the great works which have since achieved world-wide renown.

When he introduced his true planes into his own workshop, the workmen did not conceal their prejudice in favour of the old method of planing, but a very short experience taught them that the new process was much quicker and more reliable than the old one. The new planing machine may be said to have come perfect from its creator's hands, for since its first construction it has continued to be made to this day without any alteration in principle.

The development of the principle of the slide followed the construction of true planes, the one being in fact an extended application of the other. Simple as the slide now appears, its application to the construction of machines to an extent that has not found a limit has only taken place during the last forty years, for though the principle itself merely consists in so forming two adjacent plates as parts of a machine that one of them can slide freely upon the other, it required years of skill and ingenuity to work out its application in complicated machines, and its effect in facilitating the operations of the workshop might be regarded as next in importance to the introduction of the steam-engine as a source of motive power. It gave birth to a better and happier age.

The improvement of the screw was another important step in the same direction. He observed that great inconvenience was experienced owing to the variety of threads adopted by different manufacturers in the bolts and screws used in fitting up steam engines and other machines, and that the work of repairing was thus rendered both expensive and imperfect. He saw that the evil would be remedied by a uniform system in which the threads would be constantly the same for a given diameter. The old diversity of threads made it necessary in the refitting shop of a railway or shipbuilding company to have as great a variety of screwing apparatus as there were different manufacturers by whom the engines were supplied. It appeared obvious to him that if the same system of screw threads were in common use, a single set of screwing machines would suffice. He therefore directed his attention to this matter. He was led to alter the threads of the screws used at his own works in consequence of various objections that were urged against them. An extensive collection of screw bolts was made from the principal workshops throughout England, and the average thread was carefully observed for the different diameters. The $\frac{1}{4}$, $\frac{1}{2}$, 1, and $1\frac{1}{2}$ inch were particularly selected, and taken as fixed points of the scale by which the intermediate sizes were regulated. The scale was afterwards extended to six inches. Impressed with the necessity of improving the guiding screw in the lathe, he determined in like manner to accomplish this desideratum. He got rid of fractional measurements, and made the screw as perfect as possible. In the guiding screw, thirty feet long, there were two threads in the inch, and he worked upon it every day for six months, making it take out its own errors. As then perfected it has remained ever since, and the standard thus created is one of the most important as well as one of the nicest applications of mechanism. It took a long time to do it, but the result is that now every marine engine and every locomotive in this country has the same screw for every given diameter. His system of screws has now been adopted throughout the world, wherever engines and machinery are manufactured, the dies for producing the whole series having been originally furnished from his works at Manchester.

By his multiform applications of the true plane, the slide, and the screw, he enabled mechanics to work with a facility, precision, and cheapness hitherto unknown ; but to insure the perfection of such work, exact uniformity of manufacture and reliable means of testing this exactness were still wanted. "I cannot impress too strongly on this institution," said Sir Joseph to the Institution

of Mechanical Engineers in 1856, "and upon all in any way connected with mechanism, the vast importance of possessing a true
plane as a standard of reference. All excellence of workmanship
depends upon it. Next in importance to the true plane is the
power of measurement." His measuring machine supplied that
power. It is one of his simplest and yet most valuable inventions.
The principle of it is best described in his own words : "The
measuring machines which I have constructed are based upon the
production of the true plane. Measures of length are obtained
either by line or end measurement. The English standard yard
is represented by two lines drawn across two gold studs sunk in
a bronze bar about thirty-eight inches long, the temperature being
about 62° Fahr. There is an insurmountable difficulty in converting line measure to end measure, and therefore it is most desirable for all standards of linear measure to be end measure.
Line measure depends on sight aided by magnifying glasses ; but
the accuracy of end measure is due to the sense of touch, and the
delicacy of that sense is indicated by means of a mechanical
multiplier. In the case of the workshop measuring machine the
divisions on the micrometer wheel represent 10,000ths of an inch.
The screw has 20 threads to an inch, and the wheel is divided
into 500, which multiplied by 20 gives for each division the
10,000th of an inch. We find in practice that the movement of
the fourth part of a division, being the 40,000th of an inch, is
distinctly felt and gauged. In the case of the millionth machine,
we introduce a feeling piece between one end of the bar to be
measured and one end of the machine, and the movement of the
micrometer wheel through one division, which is the millionth of
an inch, is sufficient gravity. The screw in the machine has 20
threads, which number multiplied by 200—the number of teeth in
the screw wheel—gives for the turn of the micrometer wheel the
4,000th of an inch, which multiplied by 250—the number of
divisions on the micrometer wheel—gives for each division one

Q

millionth of an inch. The sides of this feeling piece are true planes parallel to each other, and the ends, both of the bars and the machine, are true planes parallel to each other and at right angles to the axis of the bar; thus four true planes act in concert. In practice we find that the temperature of the body exercises an important influence when dealing with such minute differences, and practically it is impossible to handle the pieces of metal without raising the temperature beyond 60°. I am of opinion that the proper temperature should be approaching that of the human body, and I propose that 85° Fahr. should be adopted, and that the standards and measuring appliances should be kept in a room at a uniform temperature of 85° Fahr." Again he states that when a standard yard, which is a square bar of steel, is placed in one of his large measuring machines, and the gravity piece is adjusted so as to fall by its weight, the heat imparted from the slightest touch of the finger instantly prevents its fall, owing to the lengthening of the bar by so small an amount of heat. As another illustration of the extremely minute quantity represented by the millionth part of an inch, he says it is only necessary to rub a piece of soft steel a very few times to diminish its thickness by a millionth of an inch.

Another striking illustration of the utility of exact measurements and the importance of very small differences of size was given by him in an address delivered at Manchester in 1857. " Here," he said, " is an internal gauge, having a cylindrical aperture ·5770 inch diameter, and here also are two external gauges or solid cylinders, one being ·5769 inch, and the other ·5770 inch diameter. The latter is one ten thousandth of an inch larger than the former, and fits tightly in the internal gauge when both are clean and dry, while the smaller ·5769 gauge is so loose in it as to appear not to fit at all. These gauges are finished with great care, and are made true after being case hardened. They are so hard that nothing but the diamond will cut them, except the grinding process to which they have been subjected. The effect of applying a drop of fine oil to the surface of this gauge is very remarkable. It will be observed that the fit of the larger cylinder becomes more easy, while that of the smaller becomes more tight. These results show the necessity of proper lubrication. The external and internal gauges are so near in size that the one does not go through the other when dried, and if pressed in there would be a danger of the surface particles of the one becoming imbedded in or among those of the other, which I have seen happen ; and then no amount of force will separate them ; but with a small quantity of oil on their surface they move easily and smoothly. In the case of the external gauge, ·5769 in diameter, which is one ten thousandth of an inch smaller in diameter than the internal gauge, a space of half that quantity is left between the surfaces, this becomes filled with oil, and hence the tighter fitting which is experienced. It is thus obvious to the eye and the touch that the difference between these cylinders of one ten thousandth of an inch is an appreciable and important quantity, and what is now required is a method which shall express systematically and without confusion a scale applicable to such minute differences of measurement." The Whitworth gauges have been adopted by the Government as standards of measurement.

In 1842 he produced a simple machine which at the time did more to bring his name under the immediate notice of the general public than all his wonderful tools. This was a street sweeping machine, which after being in operation ten months in Manchester was reported to have changed that town from being one of the dirtiest into one of the cleanest of our large towns. A contemporary account says the power of the machine was extraordinary, being equal to thirty men, while in its operation the numerous annoyances inseparable from the old hand method were avoided. The apparatus consisted of a series of brooms suspended from a light frame of wrought iron hung at the end of a common cart, the body of which was placed as near the ground as possible for greater facility of loading. As the cart and wheels revolved, the brooms successively swept the surface of the ground, and carried the soil up an inclined plane, at the top of which it fell into the body of the cart.

It would be tedious to go into all his minute inventions, but his activity may be indicated by the fact that between 1834 and 1849 he took out fifteen patents, being an average of one per annum. This was the period in which most of his important machine tools were brought to perfection. At that time a patent cost from 500*l.* to 600*l.*, so that it was not worth while to patent every trifling improvement.

All the machine tools invented by Sir Joseph Whitworth were exhibited at the Great Exhibition of 1851; and the reports of the juries on the machinery departments are full of compliments to him. When we remember the alarm that was then raised at the technical and mechanical progress made by foreign nations, the decisive judgment of the jury in this department is of historic interest. In concluding their report they said: "From the above description of engineers' tools, it will be seen that Mr. Whitworth has contributed one or more specimens of first-rate excellence under each head. In addition, it is necessary to direct particular attention to his measuring machine (which, however, properly belongs to the class of philosophical instruments) and to the admirable collection of apparatus by the employment of which a uniformity of system in the dimensions and fitting of machinery and in the sizes and arrangement of screw-threads is rendered practicable amongst engineers in general. The confusion and delay occasioned in the repair of machinery and apparatus of all kinds by the want of such a system have long been felt; and the attention of engineers was directed to the subject by a paper that he communicated to the Institution of Civil Engineers in 1841. His system has already obtained great extension, and he has contributed to the present Exhibition a complete set of the apparatus required to carry it out."

The Council awarded to Sir Joseph Whitworth their medal for his engineers' machine tools, measuring machine, knitting machine, &c.

From his primary inventions a whole progeny of tools, gauges, and machines were produced for drilling, shaping, slotting, and numerous other purposes; and as the result of these manifold improvements the production of machinery has multiplied during the last forty years to an enormous extent. By the exactness and uniformity of manufacture which have been attained by his improvements, manufacturers can now supply in any number the component parts of a machine which are perfectly interchangeable. For example, fifty years ago the thousands of spindles in a cotton factory had each to be separately fitted into the bolster in which it had to work. Now all these spindles are made to gauge and are interchangeable. As an example of the efficiency attained by these improvements, take the duplex lathe, which he designed and perfected, and used almost exclusively at his own works. By means of it the prices of work were reduced 40 per cent., and in addition to that 40 per cent. more work was produced.

But the beneficial effects of these mechanical improvements, small and unimpressive as they appear individually, are not confined to the workshop. No one appeared to take a sounder view of this subject than their inventor. In the concluding passages of his presidential address to the Institution of Mechanical Engineers in 1857, Sir Joseph made some prescient observations, which were then regarded by not a few as economic heresy. " Formerly," he said, "when the wealth of the nation was produced, as it were, by hand labour, a different state of things existed to that of the present day (1857). Our means of production are now increased in some cases more than one hundred and in others more than one thousand fold, and this will go on just in proportion as the masses of the people are able to consume larger quantities of everything they may require. When the farm labourer pays less for his sugar and tea, more meat will be consumed (which again goes to improve the land), also the more he will have of our manufactures. In this wonderful power of producing wealth which now exists, none can be more interested and benefited than the proprietor of the land. A striking proof of this is given by its increased value in the manufacturing counties and for miles adjoining our manufacturing towns. The competition, too, of our manufacturers and merchants to be possessors of land is shown by the small rate of interest with which they are satisfied for the outlay of their capital on the soil. The proprietors of land may rest assured that in the future development of mechanical improvements, none will be more benefited than themselves. I don't hesitate to say that all the harvest operations of land, properly laid down, will very shortly be performed, in one quarter of the time required with the hand labour now expended, by the further application of machines worked by horse-power. This is my conviction based upon the experience I have had in the successful working of the machine I constructed for sweeping the streets and at the same time filling the cart by horse-power. By the combined aid of mechanical improvements, of the science of chemistry, together with the greater skill of our modern agriculturists, the cultivation of the land throughout Great Britain must more and more approximate that of a garden."

Such were some of the material results that he anticipated as the effect of mechanical skill. He was next to exercise his inventive powers in another field, success in which had often been attended with very different results from those just described by himself as the fruits of industrial progress. From the foremost

position as a maker of implements of construction to a similar
position as a maker of the apparatus of destruction may not
now-a-days appear a kind of natural sequence ; but in the case
of Sir Joseph Whitworth this was by no means a rapid transition.
Nor was it a voluntary transition. At the outbreak of the
Crimean War the equipment of our army became an urgent
question. The Enfield rifle was then considered the best weapon ;
and public confidence in it was strengthened by the accounts given
of its effectiveness at the battles of the Alma and Inkerman, where
"it smote the enemy like the destroying angel." But this favourite
weapon was then made by hand, and when it became necessary
to obtain rifles in large quantities it was found that the private
makers were unable to supply them in the requisite time. Under
these circumstances application was made to Parliament by the
military authorities for the means to erect a small arms factory ;
and as the Birmingham gunmakers vigorously opposed this
proposal, a select committee was appointed to consider the
subject. Amongst the witnesses examined before this committee
was Sir Joseph Whitworth, who contended that there were no
sufficient data available for their guidance in the establishment
of a small arms factory, that little or nothing was then known
of the real nature of the interior of barrels, and that the most
effective form of barrel was only determined by the vague rule
of thumb. To demonstrate the mechanical precision with which
the work could be done, he explained that it was possible to
measure sizes to the millionth part of an inch, and at the same
time pointed out that till the accuracy of the barrel was tested
by such wonderful measuring power, no sound conclusions could
be arrived at as to the best form of bore, and that the Government, in proceeding to establish a small arms factory without
clearer data of this sort, would be going to work in the dark.
His view was amply confirmed by the differences of opinion
among gunmakers as to the best form of bore ; while Lord

Hardinge, then Master-General of the Ordnance, admitted that "one rifle shoots well, another ill, and the eye of the best viewer can detect no difference in the gun to account for it." In these circumstances Lord Hardinge invited Sir Joseph in 1854 to furnish designs for a complete set of new machinery for making good rifles. This invitation he at first declined, and urged as his reason "the importance of first ascertaining what the principle of the unknown secret was that caused the differences in the power of rifles, before any machine could be constructed to make a rifle that would require no further alteration." As the result of further communications with Lord Hardinge, Sir Joseph offered to conduct a series of preliminary experiments with the view of determining the true principle governing the construction of rifle barrels, if the Government would pay the expense of erecting a shooting gallery for experimental purposes. At first this offer was not accepted, chiefly on the ground of "expensiveness;" but, after being some time in suspense, Lord Ellesmere and Lord Hardinge urged upon the Treasury the necessity of its immediate acceptance. Lord Hardinge represented that "the most celebrated mechanician of his country" had declined the responsibility of producing the required machinery until he had first, by the most careful experiments, ascertained the true principle for constructing and rifling the barrel; and so essential did he consider this precaution that Sir Joseph would rather defray the attendant expenses himself than proceed without the preliminary experiments. The Government wanted, at the earliest possible date, a million rifles, which Birmingham could not supply with the existing means of production in less than twenty years. "No gunsmith," he added, "could imitate the most perfect rifle, nor tell why it shoots well or ill; but if the secret be discovered, it may be copied by machinery, and Sir Joseph Whitworth is confident that he can discover and can copy it." Assured that if this necessary demand were denied, there would be an end to the plan of making rifle barrels by machinery, the Lords of the Treasury assented to the experiments in May, 1854. Sir Joseph then ordered the erection of a suitable gallery, protected from changes in the wind and from fluctuations in the atmosphere. He considered it absolutely necessary to track the path of the rifle bullet through its entire course, to determine whether its point preserved the true forward direction, and to mark its trajectory. This rifle gallery, erected at Rusholme in October, 1854, was accordingly furnished with screens of very tight tissue paper. Measuring 500 yards in length, 16 feet in width, and 20 feet in height, the gallery had a slated roof and was lighted by openings in the south side. It was furnished with a target on wheels, so that it could be used for shooting at various distances, and there were also rests for steadying the aim. But the experiments were delayed by an unexpected accident. The gallery had scarcely been a week finished when it was blown down by a storm of great violence, and it was not till the spring of 1855 that it was rebuilt. The experiments in it began in March. Lord Hardinge, who took a great interest in these experiments, communicated the results to the Prince Consort. The Enfield rifle had a bore of 577 inch and the rifling had one turn in seventy-eight inches. The experiments carried on by Sir Joseph led him to the conclusion that the Enfield rifle was wrong in every particular. His first step was to find out the effect of increased twist in rifling, and for this purpose he tried barrels with one turn in sixty inches, one in thirty inches, one in twenty inches, one in ten inches, one in five inches, and one in one inch, by firing from each of those shapes mechanically fitting bullets of lead and tin. In this way he satisfied himself that the best twist for a rifle bullet was one in twenty with a minimum diameter of barrel of 45 inch. In the Enfield, moreover, the bore was cylindrical with grooves, while in the new Whitworth rifle it was made hexagonal with the edges rounded.

The experiments which gave these results extended over two years, and when, at the end of that time, Sir Joseph announced that he had solved the problem, a series of trials were made with his new rifle, first at Hythe and next at Woolwich. It beat the Government musket in the proportion of three to one. The best performance previously noticed had been a deviation of twenty-four inches in 500 yards; the deviation of the Whitworth rifle at 800 yards was barely half this amount, while the deviation of the Enfield at 800 was more than twice its deviation at 500. At 1,400 yards the Whitworth deviation was 4·62 feet, while at that distance a target fourteen feet square was not hit at all by the Enfield.

Sir John Burgoyne, who took a lively interest in the inquiry, stated, in a letter dated December 11, 1857, and addressed to Sir Joseph Whitworth: " I am myself so fully satisfied with your musket as a great step in advance which I must expect to be early adopted, that I am anxious for the time when we can know more about its powers of penetration of different substances, say iron, steel, elm, oak, &c., or earth, with the hard bullet." Sir Joseph soon solved this point. His rifle sent its bullets through thirty-four half-inch elm boards, while the Enfield penetrated twelve of the boards and was stopped at the thirteenth. A new capacity of the Whitworth rifle was at the same time manifested. The rifling of the new weapon permitted the use of a steel bullet, which pierced plates of iron half an inch thick, not only point blank, but at obliquities varying from 0° to over 50°. He adopted as the best form of bullet one with a conical front and a length of three or three and a half times its own diameter. " In some projectiles which I employ," he says, " the rotations are 60,000 a minute. In the motion of machinery 8,000 revolutions in a minute is extremely high, and considering the *vis viva* imparted to a projectile as represented by a velocity of rotation of 60,000 revolutions, and the velocity of

progress of 60,000 feet per minute, the mind will be prepared
to understand how the resistance of thick armour plates of
iron is overcome when such enormous velocities are brought
to a sudden standstill."

What a handsome reward the unselfish and ingenious inventor
deserved at the hands of the British Government for producing
such a powerful weapon ! How did they recognise his services ?
They first depreciated his rifle, then ignored himself, and
finally appropriated the chief principles he had discovered. A
committee of officers in 1859 gave a conflicting account of the
Whitworth rifle which, on the whole, they made to appear
unsuitable for the British army. Individual members of the
committee were emphatic in its favour. Thus, General Hay, then
considered one of the greatest authorities on the subject, reported
that the Whitworth rifle, as compared with the Enfield, possessed
great increase of precision and range ; great increase of strength
and durability, and great increase of penetration (nearly treble).
Nor was the individual testimony of great authorities the only
evidence in its favour   The National Rifle Association put forth
an advertisement in 1860 calling for the most perfect rifle that
could be invented. The competition was open to all the world,
and on trial the Whitworth was not only declared to be the
best rifle known, but it was at once adopted. In 1861, when
the advertisement was repeated, no competitor came forward
to challenge the supremacy of the Whitworth rifle. *Apropos*,
the Queen inaugurated the first prize meeting of the National
Rifle Association at Wimbledon by firing the first shot with
a Whitworth rifle on the 2nd of July, 1860. The rifle was
mounted on a mechanical rest, which was dependent for its
geometrical exactness on the use of one of the Whitworth true
planes, and which was constructed in a way that was believed
to secure accuracy of aim at the moment of firing. On the
rifle thus placed being fired by the Queen pulling a silken cord

attached to the trigger, the bullet struck the target an inch and a half from the centre.

"Considering," said Sir Hussey Vivian, "that that shot was calculated beforehand, and that it was fired in the open air at a distance of 400 yards from the target, it was probably the most marvellous shot that ever came from a rifle."

The fame of the rifle soon spread abroad. On the second Monday in October, 1860, Sir Joseph Whitworth arrived in Paris, and on Wednesday the Emperor Napoleon sent him a message that he wished to see him at St. Cloud. The interview that followed was a long one, and in the course of it the Emperor discussed with Sir Joseph numerous questions relating to the construction of his rifles, finally expressing a wish to see them shot with. Accordingly a trial was arranged, and it took place at Vincennes with the usual successful results. The Whitworth rifle was tried in competition with the best rifles that were produced in France. At 500 yards the Whitworth carried off the honours at the rate of three to one ; at 700 yards the French rifles were withdrawn from further contest, while the Whitworth continued to shoot accurately up to 1,000. Next day it was intimated to Sir Joseph that the experiment was considered satisfactory, that the Emperor wished a number of rifles to be made for him, that he would send an officer to Southport to see some of Sir Joseph's large guns tested as soon as the necessary arrangements could be made, and that he was prepared without delay to negotiate the purchase of the patent for France, if the ammunition used presented no difficulty.

In 1862 a committee appointed by the British Government reported that makers of small bore rifles having any pretence to special accuracy had copied to the letter the three main elements of success adopted by Sir Joseph Whitworth, namely— the diameter of bore, degree of spiral, and large proportion of rifling surface. It was not till 1874 that the English Government adopted his principles, and then it was not done under his name. The Martini-Henry rifle, which then became the adopted weapon, is rifled polygonally, but a heptagon is substituted for the Whitworth hexagon. The twist in the Martini-Henry rifle is one in twenty-two, or nearly the same as that of the Whitworth rifle; and the length of the bullet used for the Martini-Henry—2·93 diameters—is practically the Whitworth measurement.

In a debate in the House of Commons in 1861, Mr. Turner said that "Sir Joseph Whitworth was engaged in manufacturing instruments of peace when he was applied to by the Government. It was not his wish to enter upon the manufacture of these rifles; but having been applied to by the Government, he gave his skill and energy, and devoted thousands of pounds of his own private fortune, to carry on experiments which were first originated at the instance of the Government. Sir Joseph had no desire to make a fortune by the manufacture of rifles, but he had a desire to benefit his country by the production of the best weapon that could be manufactured for its defence, and to bring his skill to bear in producing an arm which should enable the troops of this country to meet the enemy, both at home and abroad, with effect. He ended by producing the very best weapon ever invented. That was the only honour he sought, and he has attained it, although very scant justice has been done to him since he completed his experiments."

Professor Tyndall says: "When Sir Joseph Whitworth began his experiments he was as ignorant of the rifle as Pasteur of the miscroscope when he began his immortal researches on spontaneous generation. But, like the illustrious Frenchman, Whitworth mastered his subject to an extent never previously approached. He found the powder used for rifles unfit for its purpose. In point of precision, he obtained a ' figure of merit ' greatly superior to any obtained. He carried his ranges far beyond all previous ranges; and, in point of penetration, achieved unexampled results. He did this, moreover, by a system of rifling peculiar to himself, which had never been thought of previously, and which is substantially adhered to in the favourite weapon of to-day. It would be difficult to point to an experimental investigation conducted with greater sagacity, thoroughness, and skill, and which led to more important conclusions."

"It is the province of the generalising faculty to extend to the largest phenomena the principles discerned in the smallest." Exercising this faculty, Sir Joseph Whitworth applied to the manufacture of heavy guns the same principles that he applied to his rifle. "I have always found," he says, "that what I could do with the smaller calibres of my system could be reproduced in the larger sizes."

Having, in 1856, demonstrated that a rifle bullet should be three diameters long, he showed, when he came to the construction of ordnance, that "this rule holds good for a thirty-five ton gun as well as for a rifle." He went through a similar course of experiments to show the relation of the amount of twist to the length of the projectile, and found that projectiles varying from one to seven diameters in length yielded the following results :—With one turn in ten inches, all projectiles went steadily with the point first. With one turn in twenty inches, the projectile became unsteady when more than six diameters. With one turn in thirty inches, it fell over when more than five diameters in length. With one turn in forty-five inches, the projectile turned over and flew very wild when more than three diameters in length. From these facts he concluded that unless a gun be rifled with a quick pitch, so as to give a high rotation to the projectile, it would not be possible to fire long projectiles.

In this way he produced guns of great power and precision, and their fame was soon spread abroad. In 1863 the *Times'* correspondent with the Confederate army in America stated that it was impossible to praise too highly the performances, as a field piece, of the twenty-pounder Whitworth. "There is no other gun on the Continent," he added, "which can compare with it in lightness, precision, and length of range. Again and again one single Whitworth gun has forced Federal batteries to change their position, and eventually to fall back. There has been more joy among the ordnance officers over the arrival of five more Whitworth guns, which have just safely arrived at Wilmington, than there would be at the capture of a hundred such guns as were in position on both sides at the battle of Fredericksburg." The renown of these guns became so universal that the inventor received overtures from almost every government in Europe to be supplied with them. In 1864 the British Government arranged for an elaborate series of experiments to take place at Shoeburyness for the purpose of testing the relative qualities of the Whitworth and Armstrong guns. The competition was divided into thirty-four stages, and upwards of 2,500 rounds of shot and shell were fired. The committee of artillerists, after much deliberation, reported that they could not determine which was the superior gun. Though Whitworth's ordnance then failed to be adopted by the British Government, his guns continued to be freely exported to other countries, while not a single Armstrong gun went abroad.

In 1868 Sir Joseph again astonished the world by producing a gun which could throw projectiles further than they had ever been thrown before. Its range was 11,243 yards with 250 lb. projectile, and 11,127 yards with 310 lb. projectile; in other words, a mass of 2¼ cwt. of iron was hurled a distance of 6⅜ miles.

Continuing his experiments, he demonstrated the form of projectile best adapted for flight and penetration. It is now readily conceded that the same flowing lines, which give the form to a ship to enable it to pass freely through water, or to a bird's body to glide through the air, would apply to a bullet also. But it was

Sir Joseph Whitworth who demonstrated this to be the case, showing that for flight the head should be shaped to the curve of least resistance and the rear tapered off. He showed that the form of rear is especially important for long ranges. For short ranges one is as good as the other, but for long distances a shot with a tapered end will reach one mile further than one with parallel ends. He also demonstrated, contrary to what was then generally believed, that a flat-headed projectile will penetrate armour-plates even when striking obliquely, and that is the only form that will pass in a straight line through water. From 1858 he advocated the flat front, and in 1869 he gave the following as the points of superiority, which he claimed to have established for his projectiles : That the flat-fronted form is capable of piercing armour-plates at extreme angles ; that increase in length, while adding to the efficiency of the projectile as a shell, in no way diminishes, but, on the contrary, proportionately improves its penetrative power ; that the amount of rotation he had adopted in his system of rifling is sufficient to ensure the long projectiles striking "end on," and consequently to accumulate the whole effect of the mass on the reduced area of the flat front. These results applied to his rifle and his ordnance. He also invented a cartridge which increased the range by from 15 to 20 per cent.

"Were it not," he says, "that the increased destructiveness of war must tend to shorten its duration and diminish its frequency —thus saving human life—the invention of my projectiles could hardly be justified ; but believing in the really pacific influence of the most powerful means of defence, these long projectiles I call the ' anti-war ' shell."

The following is his own summary of the results of his experiments on penetration :—

" In 1857 I proved for the first time that a ship could be penetrated below the water-line by a flat-headed rifle projectile.

" In 1860 I penetrated for the first time a 4½ in. armour-plate with an 80 lb. flat-headed solid steel projectile.

" In 1862 I penetrated for the first time a 4 in. armour-plate with a 70 lb. flat-headed steel shell, which exploded in an oak box supporting the plate.

" In 1870 I penetrated with a 9 in.-bore gun three 5 in. armour-plates interlaminated with two 5 in. layers of concrete.

" In 1872, with my 9-pounder breechloading gun and a flat-headed steel projectile, I penetrated a 3 in. armour-plate at an angle of 45°.

" All these performances were the first of their kind and were made, with one exception, with flat-headed projectiles, which alone are capable of penetrating armour-plates when impinging obliquely, and which alone can penetrate a ship below the water-line."

While the British Government were unable to appreciate these performances, Sir Joseph continued to supply foreign nations with his most powerful weapons of destruction, and happily no occasion has ever arisen for using them against his own country, where alone they were without honour. In 1871 Brazil put them to the test of experiment, and the committee intrusted with the decision of the comparative merits of the different kinds of ordnance, after considering for nearly two years the various systems of cannon, pronounced definitely in favour of the Whitworth gun, as that which, in respect of material and mode of construction, approached nearest perfection. The committee emphatically condemned the French system of cast-iron strengthened by wrought-iron bands as unscientific and practically inefficient. The Krupp gun, made of Krupp cast-steel strengthened with bands, they considered unreliable, notwithstanding its fine material, chiefly owing to the uncertainty and irregularity of effect, which, they said, always attended the action of the hammer, however ponderous, on masses of iron. They considered the Armstrong, the Woolwich,

R

and the Whitworth cannon much superior in construction and strength to the best produced on the Continent, the Woolwich being, in their opinion, an improvement on the Armstrong, and the Whitworth far ahead of both in the essential qualities of a good gun. The superiority of the Whitworth cannon they attributed to the homogeneous quality of the steel used, to the care exercised in its selection, to the use of the hydraulic press instead of the hammer, and to the mode of constructing the gun. With reference to its duration, the committee stated that the Krupp gun had an average life of 600 or 800 shots, while the Whitworth cannon used by the Brazilian forces during the Paraguan war averaged from 3,500 to 4,000 shots each, without a single case of bursting or serious damage having occurred.

Meanwhile Sir Joseph had solved another problem, which not only increased the power of his guns, but gained for him a conspicuous position in the ranks of British metallurgists. After determining the principles that should regulate the construction of guns, he next sought to improve the material of which guns were made. At an early stage of his experiments and for a long period this subject occupied his attention. His early interest in it is shown by a passage in his presidential address to the Institution of Mechanical Engineers in September, 1856. About a month after Sir Henry Bessemer's public announcement of his great invention, Sir Joseph said : " With regard to the manufacture of malleable iron and steel, it was with great gratification that I read the account of the Bessemer process, so beautiful and simple as apparently to leave nothing further to be desired in that part of the process. I need not tell you of what vast importance it must be to those who are more immediately connected with those branches of mechanics requiring nicety of workmanship to have iron and steel of a better quality. I may mention that in making rifle-barrels for the experiments I have undertaken for the Government, one of the greatest difficulties I encounter in attaining the degree of accuracy I require arises from the defects in iron. What we want is iron of great strength, free from seams, flaws, and hard places. Inferior iron (with the use of other defective and improper materials) is perhaps the main cause of one of the greatest errors committed in the construction of whatever in mechanism has to be kept in motion."

In the earlier years of his artillery experiments he found that the terribly destructive weapons he was creating required a better metal for their construction. All guns were then made of iron. It was the common belief till then that steel was an unsafe material to use for ordnance. He, however, turned his attention to the adaptability of steel for that purpose after the Bessemer process became a success ; but believing that its existing qualities would have to be improved before the prejudices of the period could be overcome and the tests required could be satisfied, he determined to effect this improvement. The Bessemer steel then made was hard enough, but it did not possess the requisite ductility and soundness. Ductility was indispensable, but it could not be got without forming air cells, which made it un-sound. On splitting open a large ingot of the best Bessemer steel, he found the upper part full of air cells. So far as he knew, the only mode then used for working steel so as to render it close, strong, and ductile was to compress it when in a solid state by expensive machinery. Convinced that this process could be improved upon, he began to make experiments with that object steadfastly in view ; but it seemed a herculean task.

In both French and Prussian steel works experiments with the same object in view were simultaneously carried on, but were abandoned as hopeless. Between 1863 and 1865 he laboured almost incessantly at this work, and in the latter year he succeeded in attaining his end, his ultimate success being the result of 2,500 experiments. His process simply consisted in subjecting the steel in its fluid state to such a high pressure that the air or gas bubbles were pressed out of it. He found that a pressure of not less than six tons to the square inch was required, and that under that pressure a column of steel was compressed in five minutes to the extent of $1\frac{1}{2}$ in. per foot of length, or one-eighth of its whole length. He had made the experiment of putting on the enormous pressure of twenty tons to the square inch, but found that a greater pressure than six or nine tons to the square inch did not improve the metal. He found too that the larger the mass of steel the more beneficial and effective was the hydraulic pressure. Experience also showed that the gases could not be expelled when the metal was in a semi-fluid state, but that it was desirable to get the pressure on as soon as possible after the metal was poured into the mould. These facts indicate what a difficult and dangerous process it was. At first he experienced many failures; and not his least difficulty, after evolving the principle of his process, was the making of materials and machinery capable of resisting such enormous pressure. He was obliged at the outset to make comparatively small presses for compressing the steel, and then gradually to increase the size of his apparatus, thus building up and extending the process till he found that it could be employed with safety and certainty in the production of articles of any required size, such as propeller shafts and cylinder linings for the largest marine engines, ordnance of large calibre, and torpedo chambers. At last he made what was known as the 8,ooo-ton press. In his own works, where the steel was made, the workmen were so afraid of the consequences that might result from the application of such enormous pressure to fluid steel, that they always ran away when the pressure was put on in order to be secure from the risk of accidents. Nor was this his only discouragement, for while the workmen were frightened, the critics were laughing at the invention. If the fluid steel were compressed in moulds of such great strength as to resist the requisite pressure, where could the air and gas bubbles escape? If, they argued, there was no way of escape, they must be still in the steel. Science was baffled to explain how it was done, but the effect of the process was nevertheless apparent. Provision was made for the escape of the gases, and there was considerable flame caused by their ignition during the time of their escape. Anyhow, the result was indisputable. When tested along with the best metals hitherto made, his was found superior to them all in strength and ductility. Damascus steel burst with a charge that compressed steel resisted. His field guns are now forged solid, and then bored and rifled, the trunnion hoops being screwed on. When Sir E. J. Reed was chief constructor for the English Admiralty, he expressed to Sir Spenser Robinson his conviction that " the Whitworth metal is superior to all other existing steels for the manufacture of ordnance. In no branch of manufacture has the want of soundness and uniformity in steel been more severely felt than in this, and in none is a superior steel more essential. No one who has considered the process of Sir Joseph Whitworth and has examined the steel produced by it can, I think, doubt for a moment that it is a more close, a more compact, and a more perfect material than any other description of steel in existence." Few would now dare to dispute the general accuracy of that opinion, but at the time it was uttered few men in high official position were so bold as to assert it. Sir Joseph's own accounts were more definite. He conducted a series of experiments on small cylinders of metal 2 in. long, and having an average section of half a square inch; these he tore asunder by hydraulic pressure, and the difference in length before and after this strain gave the elongation and ductility of the metal. The conclusion he arrived at was that, taking the tensile strength per square inch of the best iron as twenty-seven tons, with a ductility of 38 per cent., and that of ordinary cast-iron as ten tons, with a ductility of ·75 per cent.—high standards of comparison—steel compressed by his process gave from forty to seventy-two tons of strength and from 32 to 14 per cent. of ductility. Another test used by him was equally convincing. He took a small cylinder like a gun-barrel, and having firmly plugged it at both ends, he fired a charge of gunpowder within it. It took six explosions, with an aggregate quantity of 250 grains of gunpowder, to burst such a cylinder, and even then it remained in one piece. In that way it has been shown that a cylinder of compressed steel is capable of enduring forty-eight explosions of 24 oz. each of gunpowder, while a similar cast-iron cylinder burst in twenty pieces at the first discharge of 3 oz.

The practical working of the process by which this steel is produced is thus described by Professor Tyndall, who saw it done at Sir Joseph Whitworth's works : " Within a hollow steel cylinder of enormous strength were placed a series of cast-iron bars, so as to form a kind of lining. The bars were laid loosely side by side, so as to admit of the passage of a gas between them. They were also grooved, with a view of facilitating gaseous motion. The bars were coated by a porous lining of sand and other materials, through which gases could readily be driven by pressure. In the middle of the cylinder stood a core, also formed so as to permit of the escape of gas from it. A space of several inches existed between the inner core and outward sheath. A large ladle was at hand, and into this was poured the molten metal from a number of crucibles. From the ladle again the metal was poured into the annular space just referred to, filling it to the brim. Down upon the molten mass descended the plunger of a hydraulic press. On first entering it a shower of the molten metal was scattered on all sides ; but inasmuch as the distance between the annular plunger and the core on the one side and the sheath on the other was only about one tenth of an inch, the fluid metal was immediately chilled and solidified. Thus entrapped, it was subjected to pressure, which amounted eventually to about six tons per square inch.

" Doubtless gases were here dissolved in the fluid mass, and doubtless also they were mechanically entangled in it as bubbles. I figure to myself the fluid metal as an assemblage of molecules, with the intermolecular spaces in communication with the air outside. Through these spaces I believe the carbonic oxide and the air to have been forced, finding their escape through the porous core on the one side, and through the porous sheath on the other. From both core and sheath issued copious streams of gas, mainly, it would seem, in the condition of carbonic oxide flame. A considerable shortening of the fluid cylinder was the consequence of this expulsion of gases from its interior. The pressure was continued long after the gases had ceased to be ejected; for, otherwise, the contraction of the metal, on cooling, might subject it to injurious internal strains. In fact, castings have been known to be rent asunder by this contraction. By the continuance of the external pressure, every internal strain is at once responded to and satisfied, and the metal is kept compact.

" The main factors which determine the quality of any kind of steel are its strength and ductility. The method adopted by Sir Joseph Whitworth in determining these factors is, like all his mechanical contrivances, admirable. Both ends of a cylinder of a definite length and cross-section are screwed firmly between two jaws, which are then separated by hydraulic pressure. For a time this stretched cylinder maintains a uniform diameter. At a certain pressure it passes its limit of elasticity, the passage being distinctly indicated by the dial which registers the pressure. From this point forward the cylinder is observed to contract at its centre, and it finally snaps across. The 'strength' is measured by the breaking force; while the 'ductility' is determined by bringing the fractured surface close together, measuring the length of the stretched mass, and expressing its elongation as a percentage of the original length of the cylinder. In one experiment made in my presence, forty seconds sufficed to stretch and break the cylinder, and there was not the slightest jar or jerk observed during the process. I entirely sympathise with the desire entertained by Sir Joseph Whitworth, that these two elements of strength and ductility should be determined for, and registered upon, every ingot and bar of steel employed in the construction of our railway tires and axles ; and indeed on all portions of machinery, the giving way of which imperils human life. It may be added here that I have seen the fluid compressed steel in various stages of working, and sometimes in masses suitable for screw-propeller shafts or for 35-ton guns. I have never detected a flaw or fault in it, its planed surfaces always disclosing a metal of the most closely coherent texture."

In 1876 the two screw-propeller shafts of the *Inflexible* were made of steel compressed by this process. This vessel is the largest turret ship in the English navy, is protected by the thickest armour, is armed with the heaviest guns, and is, in construction and method of working, an unparalleled engine of war. Her propeller shafts were 283 feet in length and weighed 63 tons. If they had been made of wrought iron their weight would have been 97 tons, so that by using Whitworth compressed steel the constructor saved 34 tons being carried during the whole lifetime of the engines. The strength of the metal in these shafts is 40 tons to the square inch, and the ductility is 30 per cent. The shafts are 17 inches in diameter, and have a 9-inch hole through them. They were cast hollow, but with a much larger diameter, and with a considerably larger hole. In the *City of Rome*, too, which when launched in 1881 was described as a floating palace, and was the largest vessel afloat except the *Great Eastern*, the crank shaft was made of Whitworth compressed steel. It weighed 64 tons, whereas if it had been made of solid iron it would have weighed 73 tons. It is obvious that for other purposes this steel can be used with similar results. For instance, the inventor himself proposes to reduce the dead weight of railway carriages from 7 tons to 5 tons by using this steel instead of wrought iron, believing that during the lifetime of a railway carriage every ton of dead weight that can be saved represents a saving of 300*l.*

The Whitworth process for the manufacture of compressed steel was patented in 1865, but it was not till 1869 that Sir Joseph got his apparatus completed, and was in a position to manufacture his steel in quantities fit for use in his works. The patent accordingly expired in 1879, but the Judicial Committee of the Privy Council, in consideration of the cost and usefulness of the invention, agreed to prolong it for a further term of five years.

In making guns for the Brazilian navy he determined that the quality of the metal should be tested by its resistance to the explosion of gunpowder; and the results thus obtained with his breechloaders were so far superior in every way to what was possible with any muzzleloading guns that he felt sure that the British Government must ultimately adopt breechloading. Wishing to convince the British Admiralty and War Office of the superiority of his breechloading steel guns, he offered in 1874 to lend them a 7-inch gun of that description, firing 33 lbs. of powder, together with a 35 ton muzzleloading gun, capable of firing armour-piercing projectiles weighing 1,250 lbs., with a bursting charge of 58 lbs. Both offers were declined, much to the surprise of the Brazilian Government, from whom permission was obtained to lend these guns.

The resistance which fluid compressed steel offered to the bursting power of gunpowder was so clearly demonstrated by repeated trials that Sir Joseph was impressed with the idea that more powder might be fired from his breechloader with a larger powder chamber than a muzzleloader could bear. Further experiments soon proved that his breechloader could fire more than double the charge of the Government gun. In experiments made in 1876, by a commission of French officers for the

Brazilian Government, with a 35-ton gun of shorter length than the Government gun, the superiority in bursting charge was 400 per cent.

After the increase in the efficiency of weapons of destruction had been carried to such an extent, fears began to arise lest it might be found impossible to make materials capable of resisting their deadly power. Means of protection are not less necessary than means of destruction ; and to provide the latter has generally been found more easy than the former. No sooner had Sir Joseph Whitworth provided efficient ordnance than he proceeded to study the construction of armour, and in 1877 he produced his "impregnable armour plating," formed of his fluid-compressed steel, and built up in hexagonal sections, each of which was composed of a series of concentric rings around a central circular disc. These concentric rings prevented any crack in the steel from passing beyond the limits of the one in which it occured. A trial of this sort of plate made at Manchester in 1878 gave remarkable results. A small target, $2\frac{1}{2}$ inches in thickness, and representing one section of a plate, being fired at with 3 lb. shot, all the iron shot broke up against it harmlessly, and a compressed steel projectile only indented its surface to a trifling extent. The experiment was afterwards repeated on a larger scale with a target 9 inches in thickness, supported by a wood backing against a sand-bank. In front of this target a horizontal iron tube was fixed to receive the fragments of the shot. Against this a Palliser shell weighing 250 lbs. was fired from a 9-inch gun, with 50 lbs. of pebble powder, at a distance of 30 yards from the target. Such a projectile would have passed through 12 inches of ordinary iron armour-plating ; but against this new target it was powerless. It broke up into innumerable small fragments. Such was the force of impact that the target was driven back 18 inches into the sand. The fragments of the projectile escaped at the end of the tube, and continuing their rotating movement in such a way as to cut a sort of trench through ten planks immediately in front of the displaced target, were then scattered about in a shower. The only piece of any size that survived the shock was a flattened mass, 8 lbs. in weight, which was formed from the apex of the shell, and which was left imbedded in the surface of the target, where it had made for itself an excavation of about 8 inches in diameter and $1\frac{4}{10}$ inch deep at the central or deepest part. With the exception of this shallow depression, the target was absolutely uninjured, the ring which received the shock was not cracked, and no disturbance of the back surface was produced. This was then admitted to be the greatest achievement in resisting the latest implements of destruction; and competent judges declared that this new plating would supply, not only a material that was invulnerable to any missiles employed in warfare, but a lighter armour than any in use for large ironclads.

The inventor of these remarkable guns and armour plates has publicly pointed out the persistency with which the English Government continued to manufacture guns of inferior material, and to ignore the principles of gunnery that he advocated. Indeed, in one of his publications on this subject he said that, following past precedent, he supposed his name and labours would be forgotten before his own Government would fully understand the principles he had worked out. But he has lived to see indications of a better spirit on the part of our naval and military authorities, and to see that practical vindication of his principles which from the first he declared that time would inevitably ratify. In the House of Lords on the 14th April, 1878, the Duke of Somerset expressed his conviction that wrought-iron plates were not sufficiently strong for armour, the more so as Sir Joseph Whitworth had been successful in producing a very much stronger quality of plate. Shells, he said, must now be made of stronger metal, and so also must guns. He believed that a 50- or 60-ton gun made of Whitworth metal would be able to do everything that could now be done by an 80- or 100-ton gun. Sir Joseph Whitworth's shells carried double the charge of the Woolwich shells ; and he hoped that measures would be taken to secure a fair trial, because he could not help feeling that it was very natural for the Woolwich authorities to favour their own manufacture. Lord Bury, replying for the Government, stated that a trial of different kinds of ordnance was going on at Shoeburyness. The Government were aware that Sir Joseph Whitworth had produced a steel which was much stronger in every way than any that had yet been tried ; and its use was a point that was undergoing investigation !

Incredible as it may appear, the fact is that the first naval power in the world is one of the slowest in adopting improvements in ordnance. In the House of Commons on the 18th March, 1881, the Secretary to the Admiralty (Mr. Trevelyan, the biographer of Lord Macaulay) stated that, " At this moment there was not a single heavy breechloading gun mounted in any of our ships, but by the end of next year a very substantial beginning would have been made towards arming our fleet with breechloaders. The Admiralty had been driven to the step by the fact that a high velocity was now required for the projectile ; that high velocity could only be obtained by a great length of gun ; and to load a gun over a certain length at the muzzle became impracticable under the ordinary conditions of mounting guns afloat. When the shot and the powder weighed nearly a ton together, it was no light matter to ram them down a tube 30 feet long. He added that the Government had lost nothing by waiting. In the breechloading guns which were being made at Woolwich to serve, if successful, as patterns for our future naval ordnance, it had been possible to take advantage of the various improvements introduced by the *foreign* nations which had preceded us in their use. Our new breechloading guns would be, calibre for calibre, more powerful than any which had been

produced abroad, and in point of price they would compare
favourably with the ordnance of foreign nations. Our new 9-inch
breechloader of 18 tons, which would penetrate 16 inches of iron
and 13 of composite armour, cost only two-thirds as much as the
Krupp gun of the same size and certainly not greater power."

The year 1881 also witnessed the introduction of steel for the
manufacture of ordnance at Woolwich. In the autumn of that
year it was reported with *naïveté* that "a 43 ton gun largely
consisting of steel is being constructed at the Royal Gun
Factories in the Royal Arsenal at Woolwich. One of the steel
coils, 100 feet in length, and weighing four tons, was coiled on
September 20th, with an amount of simplicity and success which
fully demonstrated the workability of modern steel. Instead of
the rigid, brittle material which steel is generally supposed to be,
the metal now produced at the gun factories is almost as elastic
and docile as wrought iron, and at a very slight sacrifice of its
carbonic virtues it can be produced in bars or plates as cheaply
as iron, and nearly as available for every purpose. These qualities
deprive it of the objections hitherto entertained with regard to
its employment in gun-making, and the 43 ton breechloader
now in progress will, with the exception of the outer coil or jacket,
be nearly all steel. In the experiments which have been carried
out with the new material, it has been found to be remarkably
pure and free from laminæ or dross, which is always present to
some extent in wrought iron, and its toughness has been satisfac-
torily proved by every kind of strain. Welded into a homogeneous
mass, it forms a cylinder of a quality which is said to have never
hitherto been equalled. A new furnace for making the steel in
bulk is in operation at the gun factories, and establishes a marvellous advance upon the only method in vogue a few years ago,
when a bar of steel required for its production some hundreds of
small crucibles."[1] These properties of steel, which the Woolwich

[1] *Times,* October 3, 1881.

authorities professed to have discovered for themselves in 1881, were just what our greatest steel inventors had urged upon them in vain for a quarter of a century.

In 1883 the Marquis of Hartington, as minister for war, made the following statement in the House of Commons : " During the last few years the increasing length and size of guns have led to a complete change of system. The great weight and the enormous charges of the new guns have caused the War Department to consider the feasibility of constructing guns wholly of steel, and the experience of the Ordnance Committee, as well as of private manufacturers, together with the results of recent experiments conducted by them, have resulted in the committee making a recommendation that ' the manufacture of wrought-iron guns should be discontinued, and that all guns in future should be made wholly of steel.' The requirements of the Navy, as the first line of defence, have been mainly considered. The great length of new guns have led to the necessity of adopting the breech-loading instead of the muzzle-loading system. The Committee of Ordnance began their experiments in 1879 with breechloading 6-inch guns. These experiments were so far satisfactory that the 6-inch gun has been and is being provided for the Navy. In 1880-81, 14 such guns were provided ; in 1881-2, 103 ; in 1882-3, 59 ; and in 1883-4, 63, making in all 239. The guns of the two latter years are of steel."

In the manufacture of projectiles, too, the military advisers of the British Government discovered that there was room for improvement. Till then their projectiles had been made of iron ; the average number produced per annum was about a quarter of a million ; and the iron used in their construction varied from 5,000 to 8,000 tons per annum. In the autumn of 1881 it was announced that some valuable experience as to the construction of heavy shot and shell for the penetration of iron-clad ships was acquired by the officers of the Royal Arsenal, in a series of experiments at Shoeburyness on the 1st of October. A target had been erected representing a section of the warship most in vogue, consisting of a sound plate of wrought iron twelve inches thick, with the usual timber backing and supports; and against this was brought a 9-inch gun of twelve tons—a weapon which might be regarded as a sixth-rate gun. Being planted within sixty yards of the target, it was supposed to realise the conditions of a naval attack at close quarters. According to the account published of these experiments, it had recently been surmised that the elongated projectile, which had created such an advance in the art of gunnery since it had superseded the ancient spherical shot, might be improved as a missile of penetration by an alteration of its shape and also by a careful study of the metal composing it. " In the latter respect steel had been favoured, and the authorities had proved it capable of doing remarkable work; but the Woolwich officials had several objections to its employment in the making of shells; and they had at last produced a description of wrought iron, the precise nature of which was a secret, but the superiority of which over any material yet discovered for the purpose was said to be remarkable. It possessed the essential qualities of hardness and toughness in an extraordinary degree—a combination which had previously been considered impracticable." The trials were described as most successful, two of the shells penetrated the plate; and then broke into fragments; whereupon the public were told that a discovery calculated to double the power of the national artillery afloat could scarcely be too highly estimated.[1]

Such was the sort of language in which the world was informed

[1] It was in 1881, too, when the Government adopted various reforms in the manufacture of ordnance which Sir Joseph Whitworth had so long advocated, that they also adopted his standard gauges, increasing by from quarters of an inch to one-thousandth of an inch, as Board of Trade Standards, in like manner as if they were mentioned in the " Weights and Measures Act, 1878."

that the artillery advisers of the British Government had "discovered" a projectile which, fired from a 9-inch gun, was capable of penetrating a 12-inch iron plate. Eleven years previously Sir Joseph Whitworth penetrated, with a 9-inch-bore gun, fifteen inches of armour plating and ten inches of iron concrete!

In 1883 a new 21-ton gun made in Sir Joseph Whitworth's works at Manchester, for the Brazilian Government, gave still greater results. In the trials for range the gun was loaded with a projectile weighing 300 lbs., and a charge of 181 lbs. of powder. At an elevation of ten degrees, a velocity of 1,990 ft. per second and a range of 7,876 yards were obtained; and it was estimated that a full charge (200 lbs.) would at the same elevation carry the shot upwards of five miles. A special target was erected in order to test its power of penetration. This target consisted of a solid wrought-iron plate 18 in. thick, with a backing composed of a steel hoop 37 in. long, with a 23-in. hole, and rammed hard with wet sand, then a second backing composed of T-iron riveted on to a steel plate $1\frac{1}{8}$ in. thick, and built in solidly with oak; this was further supported by a cast-iron bed plate 20 ft. long by 5 ft. wide and $14\frac{1}{2}$ in. deep, and finally securely strutted by a series of timbers driven firmly into the sand. The gun was loaded with a Whitworth steel shell weighing 403 lbs. and a charge of 200 lbs. of powder, and was fired at about 90 ft. distance from the target. The shot, which was 9 in. diameter, went clean through the plate, about an inch from the centre, next through the hoop, bursting it open, then after passing through the second plate, it broke the iron bed plate into fragments, and finally lodged in the sand below. The shell, though slightly shortened, was found in almost perfect condition imbedded in the sand at a distance of about 17 ft. 6 in. from the point of first contact with the target.

There is another subject with which the name of Sir Joseph

Whitworth will be for ever associated. In 1868 he informed the Prime Minister that he intended to found a series of scholarships for the encouragement of young men in scientific and technical education. This announcement was as opportune as it was munificent. At that time the leading men in educational and scientific matters were deploring England's deficiency in technical education. The subject was brought prominently under public notice in 1867, when the Universal Exhibition was held in Paris. A school inquiry commission was then sitting in London, and to the Chairman of it, Sir Lyon Playfair, who had acted as a juror not only in the exhibition of that year but also in those of 1851 and 1862, addressed a letter on the advance made in England in those acquirements that come under the name of technical education. He said: "I am sorry to say that, with very few exceptions, a singular accordance of opinion prevailed that our country had shown little inventiveness and made but little progress in the peaceful arts of industry since 1862. Deficient representation in some of the industries might have accounted for this judgment against us; but when we find that of ninety classes there are scarcely a dozen in which pre-eminence is unhesitatingly awarded to us, this plea must be abandoned. My own opinion is worthy only of the confidence which might be supposed to attach to my knowledge of the chemical arts; but when I found some of our chief mechanical and civil engineers lamenting the want of progress in their industries, and pointing to the wonderful advances which other nations are making; when I found our chemical and even textile manufacturers uttering similar complaints, I naturally devoted attention to eliciting their views as to the causes. These causes are believed to be of two very different kinds. In the first place, the system of trades unionism tends to keep down the best hands by encouraging equal wages for all alike, without giving free scope to the skill and ability of individual workmen. In the second place, there was a

s

general admission that England is deficient in one advantage
which France, Belgium, Prussia, Austria, and Switzerland possess,
viz., a good system of industrial education for the masters and
managers of factories and workshops. Austria is said to possess
the best system for workmen, while the higher instruction of
managers and foremen is best attended to in France, Prussia, and
Switzerland."

This letter was considered so important by the Commission
that it was circulated for the purpose of eliciting the opinions of
eminent men of science and large manufacturers. The opinions
thus obtained were almost unanimously to the same effect. Professor Tyndall, who succeeded Professor Faraday at the Royal
Institution, stated that he had long been of opinion that "in
virtue of the better education provided by Continental nations,
England must one day—and that no distant one—find herself
outstripped by those nations both in the arts of peace and war ;
as surely as knowledge is power, this must be the result." Similar
evidence was given by Dr. Frankland, Professor of Chemistry
in the Royal School of Mines, and by Mr. Warrington Smith,
Lecturer on Mining and Mineralogy at the same institution. The
former said that in England masters and foremen rarely had any
opportunities of making themselves acquainted with the funda-
mental laws and principles of physics and chemistry ; they therefore found themselves engaged in pursuits for which their previous
education had afforded them no preparation ; and hence their
inability to originate inventions and improvements. Mr. W. Smith
believed that the greater proportional advancement made by
France, Prussia, and Belgium in mining, colliery working, and
metallurgy was due, not to the workmen, but in great part to the
superior training and attention to the general knowledge of this
subject observable among the managers and sub-officers of the
works, and that no candid person could deny that they were far
better educated as a rule than those who held similar positions in

England. Mr. E. W. Cooke, the Royal Academician, stated that at the Paris Exhibition he was struck with the great advance which Continental nations had made in ten years in the design as well as execution of works in which he had sanguinely hoped that England would have greatly excelled, if not triumphed over, our Continental neighbours. Mr. Scott Russell and Mr. McConnell were equally strong in their expressions of opinion as to England's position in mechanical art and science. "I am firmly convinced," said the latter gentleman, one of the greatest authorities on loco-motive construction, "that our former superiority no longer exists, either in material or workmanship; in fact, there are engines made in France and Germany equal to those of the best English makers. It requires no skill to predict that, unless we adopt a system of technical education for our workmen in this country, we shall soon not even hold our own in the cheapness of cost and the excellence of quality of our mechanical productions." He attributed the position of our Continental neighbours to the establishment of workmen's schools, in which a clever mechanic could qualify himself for any position in his business. Dr. James Young, the founder of the manufacture of paraffin-oil, spoke from personal experience of the value of technical education. He said : "Originally I was a working man, but have succeeded in increasing the range of manufacturing industry. The foundation of my success consisted in my having been fortunately attached to the laboratory of the Andersonian University in Glasgow, where I learned chemistry under Graham, and natural philosophy and other subjects under the respective professors. This knowledge gave me the power of improving the chemical manufactures, into which I afterwards passed as a servant, and ultimately led to my being the founder of a new branch of industry, and owner of the largest chemical works in the kingdom. It would be most un-grateful of me if I did not recognise the importance of scientific and technical education in improving and advancing manufactures."

This distinguished witness, after being present at the Paris Exhibition of 1867, said : " So formidable did the rate of progress of other nations appear to many of us that several meetings of jurors, exhibitors, and others took place at the Louvre Hotel on the subject. The universal impression at these meetings was that the rate of progress of foreign nations, in the larger number of our staple industries, was much greater than our own."

It was at the time when public interest was much exercised by these representations that Sir Joseph Whitworth announced his intention of giving 100,000*l.* to found a series of scholarships in aid of advanced scientific instruction. He sent the following letter to the Prime Minister :

MANCHESTER, 18*th March*, 1868.
28, PALL MALL.

SIR,

I desire to promote the engineering and mechanical industry of this country by founding thirty scholarships of the annual value of 100*l.* each, to be applied for the further instruction of young men, natives of the United Kingdom, selected by open competition for their intelligence and proficiency in the theory and practice of mechanics and its cognate sciences.

I propose that these scholarships should be tenable on conditions to be defined by a deed of trust regulating the administration of the endowment fund during my life, and that thereafter the management of this fund, subject to the conditions specified therein, should vest in the Lord President of the Council or other minister of public instruction for the time being.

I venture to make this communication to you in the hope that means may be found for bringing science and industry into closer relation with each other than at present obtains in this country.

I am, &c.

JOSEPH WHITWORTH.

To the Right Hon. B. DISRAELI, M.P.

The offer was of course gratefully accepted, and as an expression of public appreciation a baronetcy was afterwards conferred on the donor.

In another letter written six weeks afterward to the Education Department, he said : " I would beg leave to ask the Lords of the Committee of Council on Education to undertake the examinations for these scholarships.

" As respects the preparation of the necessary details for the examinations in the use of tools, I am willing to be responsible myself with the aid of friends, and I propose to obtain the consent of a few gentlemen to advise with me from time to time in whatever may arise in the future for my consideration.

" In reply to the invitation of their lordships to submit any suggestions, I venture to submit for consideration whether honours in the nature of degrees might not be conferred by some competent authority on successful students each year, thus creating a faculty of industry analogous to the existing faculties of divinity, law, and medicine. I am of opinion that such honours would be a great incentive to exertion, and would tend greatly to promote the object in view.

" I venture further to express a hope that the Government will provide the necessary funds for endowing a sufficient number of professors of mechanics throughout the United Kingdom."

For the first year, 1869, ten of these scholarships were offered for competition, to be held for three years on condition that the successful candidates should spend their time of holding the scholarships in the study and practice of mechanical engineering, and that the Education Department should decide on the manner of testing the scholars' progress from year to year. In order that scholars should have the utmost latitude for following the bent of their own minds, students wishing to complete their general education were permitted to go to the universities or to colleges where scientific or technical instruction was given, or to travel abroad for the same purpose. The desire of the founder was that the successful artisan should be encouraged to study theory, while the successful competitor in theory should be aided in getting admission to machine shops and other practical establishments, and accordingly it was agreed to give the same number of marks for equal merit, whether shown in theoretical science or practical skill, in the following subjects : (1) mathematics, elementary and higher ; (2) mechanics, theoretical and applied; (3) practical, plane, and descriptive geometry ; (4) mechanical and freehand drawing ; (5) physics ; (6) chemistry and metallurgy ; and (7) such handicraft processes as smiths' work, turning, filing, fitting, pattern making, and moulding.

So impressed was he with the backward condition of technical education that he proposed to give competitors the benefit of twelve months' training preparatory to the examination. With that view he offered sixty exhibitions or premiums of 25*l.* each, which were awarded for one year to young men under twenty-two years of age who undertook to compete for the 100*l.* scholarships in May, 1869 ; and in order that the benefit of these exhibitions might be widespread, they were placed at the disposal of a large number of educational bodies in different parts, such as the Universities of Oxford, Cambridge, London, Dublin, Edinburgh, Glasgow, Aberdeen, Durham, and St. Andrews, the Queen's Colleges in Ireland, King's College and University College, London,the College of Preceptors, the Science and Art Department, and the public schools of Eton, Rugby, Harrow, Westminster, Winchester, St. Paul's, Charterhouse, Merchant Taylor's, Christ's Hospital, Shrewsbury, Marlborough, Cheltenham, Manchester, Liverpool, Chester, Clifton, and Brighton ; while exhibitions for artisans were transmitted to the Society of Arts and to the Corporations of Birmingham, Bristol, Cardiff, Swansea, Halifax, Huddersfield, Leeds, Sheffield, and Northampton.

It was under these encouraging circumstances that the first

examinations were held for the scholarships in 1869 at Manchester and London. There were 106 candidates, of whom 55 had had the benefit of the year's exhibitions, while 51 had not. The results of the examination showed how low scientific education had fallen in this country. Fifty-four of the candidates failed to pass the humble standard of general scientific knowledge which had been established as essential. The remaining 52 candidates who passed the preliminary examination went to Manchester to undergo the technical examination which Sir Joseph Whitworth personally superintended; but the number that was successful at this stage was little more than the number of scholarships offered—ten. The exhibitioners passed more creditably than the others, but it was found that on the whole those who passed well in science generally failed in technical manipulation, while those who excelled as handicraftsmen were deficient in science.

Again in 1869 he provided eighty similar exhibitions to enable students to prepare for the examination for scholarships in 1870. These were, as formerly, widely distributed, six being given in London, three in Liverpool, five in Manchester, three in Glasgow, three in Edinburgh, two in Bradford, Bristol, Leeds, and Nottingham, and one in each of the other thirty-two large industrial towns in the United Kingdom. With the view of effectually promoting manual as well as scientific attainments, the scholarships were, so far as impartiality would admit, about equally divided among students of science and practical workmen. So careful and minute were the arrangements made under the fostering care of the thoughtful donor, that poor candidates competing for the scholarships were provided with travelling expenses to, and the cost of lodgings in, London and Manchester.

In 1873 Sir Joseph made a fresh suggestion to the Education Department. He said the experience of past competitions for the scholarships had proved the necessity of establishing rules which would insure that the holders of them should devote themselves to the study and practice necessary for mechanical engineering during the tenure of the scholarships. He therefore proposed that every candidate for a scholarship should produce a certificate showing that he had worked in a mechanical engineering shop, or in the drawing office of such a shop, for two consecutive years. In order, however, that this condition might not inflict any hardship on the candidates preparing for the examination of 1873, he accepted such a certificate for nine months instead of two years. Nor did his indulgence in the application of the new rule end here. Towards the end of 1873 he announced his desire that candidates for his scholarships in 1874, who, owing to shortness of notice, might not have been able to be in a mechanical shop for six months before the competition took place, should be allowed to compete, but if they were successful their scholarships should not begin until they had worked six months in a machine shop. He also suggested that the same privilege should be allowed in 1875 to candidates who had not served eighteen months in a machine shop, the scholarships not beginning till that period was complete.

At the same time he announced, with the approval of the Education Department, that the number of scholarships in the competition of 1874 would be reduced from ten to six. Each scholarship would be worth 100*l.* a year, together with an additional sum determined by the results of the progress made in the preceding year. At the end of each year's tenure of the scholarships the scholars were to be examined in theory and in practice in the same manner as in the competition for the scholarships. On the results of this examination the following payments, in addition to the 100*l.* a year, would be made among each year's scholars. To the best scholar in the examination, 100*l.* ; to the second, 60*l.*; to the third, 50*l.* ; to the fourth, 40*l.* ; to the fifth, 30*l.*; and to the sixth, 20*l.* ; provided that each scholar had made such progress as was satisfactory to the Department of Science and Art, which would determine whether the sum named, or any other sum, should be awarded. Moreover, he provided that at the expiration of the three years' tenure of the scholarships a further sum of 300*l.* would be awarded in sums of 200*l.* and 100*l.* to the two scholars of each year who did best during their term of scholarship. In this way it was made possible for the best of the scholars at the end of his period of scholarship to have obtained 800*l.*, and the others in proportion.

Sir Joseph continued to take an active interest in the management of his scheme for the advancement of technical education, and varied the conditions of examination for, and tenure of, the scholarships from time to time as circumstances required or experience suggested. Among the subsequent changes the most notable was made in compliance with the following letter which he wrote to the Secretary of the Science and Art Department on the 13th of December, 1878 :—

" The experience gained during the last ten years in the working of the Whitworth scholarships leads me to the conclusion that considerable alteration is necessary in the conditions of their tenure to secure their fulfilling the object I had in view in founding them. The withdrawal of a student for three years from his business, entailing a complete severance from its routine, seems specially to militate against the success of the scheme as a means for improving the technical education of mechanical engineers in this country. Having discussed the subject with several of my friends who are in a position to understand its bearings, I placed myself in communication with Lieut.-Col. Donnelly. The alterations which he has embodied in the revised prospectus on this head, and with regard to other points which I have considered with him, entirely carry out my wishes. I trust, therefore, that the Lords of the Committee of Council on Education will sanction these alterations being made. They may, and probably will, entail an increased expenditure during the year 1879, which would, however, be adjusted in the following year. As I understand that the excess cannot be defrayed from public funds, I request that you will inform their lordships that I shall be happy to advance what is necessary in 1879."

In accordance with this letter, the condition requiring a Whitworth scholar to devote his entire time to the prosecution of his education as a mechanical engineer was relaxed in 1880, and thereafter he had to spend at least six consecutive months in each of the three consecutive years at handicraft, not less than two of these being at the vice or lathe. The examinations in practical workmanship continued to be held in the workshops of Sir Joseph Whitworth at Manchester, which were always open to students who wished to practise the working details of mechanical engineering.

He also endeavoured to make the prosperity of his works a source of benefit to his workmen. Writing on this subject in 1877, he said : "The relations between foreman engineers and their employers have lately become of a much more intimate character, particularly in concerns which have availed themselves of the Limited Liability Act. The foreman engineers have themselves become employers. Three years since (on the 31st of March, 1874) I converted my business into that of a company under the Limited Liability Act, but not in the usual sense of asking the public to take shares in it. Myself, the foremen, and others in the concern, twenty-three in number, hold ninety-two per cent. of the shares and have practical control, while the remaining eight per cent. of the shares are held by others. In this transaction there was no good will to be charged for, and the plant was taken at a low valuation. The shares of 25*l*. were offered, as many as could be taken, to the foremen, draughtsmen, clerks, and workmen. For the workman who has not the means to buy shares, arrangements have been made that will, I think, solve some of the difficulties between capital and labour. When a workman who intends to save receives his wages, he deposits with the clerk appointed what he thinks fit. This money is employed in the concern as capital, and whatever dividend is paid to the shareholders the workman is paid for his deposits in the shape of interest on them. It has been said that these terms are more favourable to the workmen than to the shareholders; but the shareholder provides only capital, and as the workman devotes both his labour and capital the terms ought to be more favourable. If a workman from sickness or other cause wants to withdraw what he has deposited, he can by giving three days' notice receive a quarter, six days' notice a half, and twelve days' notice the whole of what stands to his credit. When a workman leaves he must withdraw his deposit; and if he holds shares he must sell them to the company at the price he paid for them." At that time the ground on which his works stood was considered worth a quarter of a million sterling.

Shortly after the announcement of this scheme Thomas Carlyle wrote the following characteristic letter to Sir Joseph Whitworth:—

"I have heard of your offer on behalf of the thrifty workpeople of Darley, and of the thankful acceptance of it by the district authorities of the place. I cannot resist the highly unwonted desire that has risen in me to say that I highly approve and applaud the ideas you have on the subject, and to declare in words that in my opinion nothing wiser, more beneficent, or worthy of your distinguished place as a master of workers has come before me for many a year. Would to Heaven that all or many of the captains of industry in England had a soul in them such as yours, and could do as you have done or could still further cooperate with you in works and plans to the like effect. The look of England is to me at this moment abundantly ominous. The question of capital and labour growing ever more anarchic, insoluble altogether by the notions hitherto applied to it, is pretty certain to issue in petroleum one day, unless some other gospel than that of the 'dismal science' come to illuminate it. Two things are pretty sure to me; the first is that capital and labour never can or will agree together till they both first of all decide on doing their work faithfully throughout, and like men of conscience and honour, whose highest aim is to behave like faithful citizens of this universe, and obey the eternal commandment of Almighty God who made them. The second thing is that a sadder object than either that of the coal strike or any considerable strike is the fact that, loosely speaking, all England has decided that the profitablest way is to do its work ill, slimly, swiftly, and mendaciously. What a contrast between now, and say only one hundred years ago! At that latter date, or still more conspicuously for ages before that, all England awoke to its work with an invocation to the Eternal Maker to bless them in their day's labour and help them to do it well. Now all England, shopkeepers, workmen, all manner of competing labourers, awaken as if with an unspoken but heartfelt prayer to Beelzebub, ' Oh, help us, thou great lord of shoddy, adulteration, and malfeasance, to do our work with a maximum of slimness, swiftness, profit, and mendacity; for the devil's sake, Amen.' "

Sir C. H. Gregory, the renter warden of the Turners' Company, and late president of the Institution of Civil Engineers, in presenting the freedom of the Turners' Company to Sir Joseph Whitworth in 1875, said : " Well has he merited fortune, fame, and honour. Raised by his Sovereign to rank and title, he has been honoured by other men of science with the distinctions of D.C.L and F.R.S. He has devoted a noble share of his well-earned fortune in munificent endowments for the higher education of mechanical engineers. When he is taken from us, he will leave his monument in the workshops of the world; and as monks of old sang requiems over the graves of departed heroes, so young mechanics, trained by his liberality, will keep the name of Sir Joseph Whitworth green in their grateful memory for all time."

SIR JOHN BROWN.

CHAPTER IX.

"O Heaven! that one might read the book of fate
And see the revolution of the times.
There is a history in all men's lives,
Figuring the nature of the times deceas'd."—SHAKESPEARE.

"SIR JOHN BROWN, my next door neighbour, was the first man to look into it," said Sir Henry Bessemer, in giving an account of the difficulties he experienced in getting steel makers to adopt his process; and thus it came to pass that, although not pre-eminently distinguished as an inventor, Sir John Brown occupies an honoured place in the history of the age of steel. He can only be classed as a nebulous star in the brilliant constellation of genius whose great inventions have permanently benefited the industrial world, but like them he was the architect of his own fortune; and his life has therefore the twofold interest that attaches to one who took a foremost part in fostering from its infancy a new industry, and who at the same time "waged with fortune a successful war." To ordinary readers such a life may have more congenial traits than if he stood upon a higher pedestal in the Temple of Science. The love of fortune actuates most men, more or less; but it is only given to the few to rise on "the wings of thought" to those commanding heights where the apple of fortune may be plucked in a moment of inspiration. Hence, to many the successful inventor is rather an object of wonder than a subject of imitation. To the general public it appears as if

> " His generous mind the fair ideas drew
> Of fame and honour which in dangers lay ;
> Where wealth, like fruit, on precipices grew,
> Not to be gathered but by birds of prey."

If the life of Sir John Brown has less of the magic spell of genius, it has more affinity with the ordinary battle of life. It displays in the unheroic arena of the workshop the application of that "conformation of the faculties" which appears to command success. The saying is as old as Demosthenes, that as it is a maxim for a general to lead his army, so a wise man should lead things and make them execute his will, instead of being himself obliged to follow events. This maxim is illustrated by the career of Sir John Brown. At first sight his connection with the Bessemer converter might appear the result of a lucky accident rather than the work of a "leading" mind ; but while, like all the creators of the age of steel, he became a man of fortune, this was not the only incident that a superficial observer might regard as an accident. From such a point of view his whole life would appear a chapter of lucky accidents. But the record of it will rather exhibit him as an illustration of Lord Bacon's observation in his *Precepts for Rising in Life ;* namely, that "there are two different kinds of men held capable of the management of affairs ; some know how to make advantageous use of opportunities, but contrive or project nothing of themselves ; whilst others are wholly intent upon forming schemes, and neglect to take advantage of opportunities as they occur ; but either of these faculties is quite lame without the other."

Sheffield, the scene of Sir John Brown's life-long labours, has for centuries been celebrated as the centre of the steel trade. Chaucer states that the gay miller of Trumpington wore a Sheffield whittle (knife) in his hose; and the distinction which the town thus appears to have attained nearly 500 years ago it retains still. But it has retained its eminence at the expense of its beauty. A popular novelist has described it as "the infernal city, whose water is blacking, and whose air is coal;" and, as is the case with most fictions, that description has some foundation in fact. But it has not been always so. In the year 1847, when Sir John Brown was "projecting" the Atlas steel works, which have since become famous, Charles Knight visited Sheffield, and addressing the members of the Athenæum, said: "It is a real good—it is to some minds a compensation for the absence of many common blessings —to live surrounded by fine natural scenery. I have heard that there is not a street in Sheffield from which you may not get some prospect of the country. The distant hills and streams are here for ever wooing the busy man to come amongst them, and receive their peace into his heart. One who was keenly alive to those influences—Ebenezer Rhodes, the topographer of Yorkshire and Derbyshire—tells us in allusion to Rome's boast of its seven hills, that Sheffield has seventy times seven, with woods, and verdant slopes, and sparkling streams innumerous. It was in these scenes that Chantrey was formed. His characteristic excellence was the union of refined taste with strong judgment. His sketches of these, his native localities, were as true and at the same time as tasteful as his statues and his busts. Think ye not that the mind of the milk boy who raised himself to equal companionship with the greatest in rank and intellect, and who, making his fortune by art, left the most splendid benefactions for the support of art of any man in any time—think ye not that the mind of Francis Chantrey grew the more luxuriantly amid the beauties of this his early home?

> " ' When calmly seated on his pannier'd ass,
> Where travellers hear the steel hiss as they pass,
> A milk boy, sheltering from the transient storm,
> Chalked, on the grinders' wall, an infant's form;
> Young Chantrey smiled.'

"The milk boy here became a mechanic—a carver. Sheffield nourished him into an artist. He drew his ideal from your scenery—his unerring tact from your practical good sense.

"Men of Sheffield! You are great travellers. Your fathers were travellers in the days when they carried their wares upon pack-horses to city and to port. You now cross the Atlantic with greater facility than those who went before you voyaged from Hull to London. You see much of manners differing widely from your own. Your minds are expanded by this familiarity with the outward world. The knowledge which you thus acquire by observation descends imperceptibly from your own firesides to your workshops and your counting houses. 'Great men have been amongst us,' says the Patriarch of the Lakes. Great men have been amongst *you;* and it is your happiness that some of them are still amongst you. There abide here memories of science, of literature, of the arts, which we all cherish, but which you must especially cherish. There are few towns of England that can boast of two such poets as still dwell in or near Sheffield, and who have drawn their inspiration from the scenes which their descriptions have rendered dearer to you. It is not an uncommon thing for local reputations to have no national recognition. It is not so with your James Montgomery and your Ebenezer Elliott. Of the productions of the one it has been said—and said by a real poet himself, John Wilson—'they are embalmed in sincerity, and therefore shall not pass away; neither shall they moulder—not even though exposed to the air, and blow the air ever so rudely through time's mutations.' The same genial critic has spoken as emphatically of the other: 'The poor might well be proud, did they know it, that they have such a poet. Not a few of them know it now; but many will know it in future; for a muse of fire like his will yet send its illumination into dark, deep holds.' . . . . Those things which were the delight of our jocund days steal in upon the sober consolations of our waning time—bright images, tender echoes. Memory dwells among the scenes in which childhood was nourished, and youth walked fearlessly."

Such was the animated description given in 1847 of the town which up to then had been the Steelopolis of England ; and which now numbers among its most distinguished men the subject of this memoir. Sir John Brown was then entering upon a remarkable career, which gave his native town a fresh lease of its ancient title. He introduced the manufacture of iron into Sheffield. He was not only the first manufacturer that adopted the Bessemer process, but in its early years was the largest producer of Bessemer steel. As a manufacturer of railway material when railways were in their infancy, he is entitled to rank among the founders of that trade ; and when the time came for the wooden walls of England to make way for iron and steel, he introduced into this country, as well as improved, the manufacture of rolled armour plates.

John Brown was the second surviving son of Samuel Brown, a slater in Sheffield, who, although not a man of fortune, possessed in no small measure that strength of character which his son afterwards displayed to better advantage in a larger sphere. The son with whom we have to deal was born in 1816, and was educated in a humble way, evidently with no idea of the commanding position which he was subsequently to attain in his native town. His scanty education was acquired in a local school conducted by a master of that old and unacademic type which is now believed to be almost extinct, but the prototype of which Goldsmith has preserved for the wonder of succeeding generations in his Village Schoolmaster. The Sheffield pedagogue kept his school in a garret ; and when the little "hopeful" first came under his rule, his brusque manner of replying to the questions addressed to him was so out of place as to excite merriment in the scholars and to offend the dignity of "the master." One scholar, a girl three years older than John Brown, sitting on the form opposite to him, was so impressed with the hazardous

T

position of the new scholar, after addressing "the master" as "Sir" with that air of decision which leaves the hearer undecided as to whether it means contempt or respect, that she went home and told her father in piteous tones of the retribution which offended majesty was likely to exact for his innocent but untutored airs. This tender-hearted girl in course of time became the most distinguished scholar that, like the boy she once trembled for, was educated at that school. She afterwards played an important part in the movement that placed Sheffield in the front rank of English towns distinguished for the efficiency of their primary education. When she died in November, 1881, it was related of her in the local press that perhaps no part of her husband's life interested her more than that of the School Board. At the opening of the Springfield Board School she was presented with a gold key and asked to perform the opening ceremony with it. At other times she gave other evidences of her kindly solicitude for the education of the young. She took a special interest in the Truants' School at Hollow Meadows; and by her death the poor boys at that school lost a compassionate friend. It was truly said that her quiet, homely, and unostentatious life, and her kindly liberality, more especially to the poor and distressed, will long be remembered by many to whose wants she had ministered; for by numerous acts of kindness, which were never intended to be known, she endeared herself to hundreds, who therefore cherished her memory with veneration, respect, and gratitude. Such was the character to the last of Mary Schofield, better known in Sheffield as Lady Brown—the partner through a long and eventful life of the boy who excited her sympathy in the garret school. The cause of her early anxiety for this, her junior schoolfellow, soon passed away, for he quickly became one of his master's favourites, and was commended for his knowledge of the English language. To the schoolmaster's honour be it said that he eventually appreciated aright the strength of the boy's character. When young Brown had reached his fourteenth year his father talked of making him a linendraper; but to his surprise the boy scorned the idea. The determined father asked why he objected to that trade; and the only answer the boy could give was: "I never will be a linendraper." Further remonstrances had the effect, first of drawing from the boy the solemn assurance that if the father insisted upon putting him to that trade he should run away and go to sea, and next of leading the father to ask what trade his son preferred. "A merchant," was the ready answer to the last question, "I should like to be a merchant;" and the only reason he gave for his choice was, that a merchant did business with all the world—a reason which, while it persuaded him most, persuaded his father least. The boy's ambition was kindled by the sight of the large establishments belonging to merchants, and the commanding position they occupied in the world; while the more sober mind of his father was appalled at the very idea of raising his son to such a princely position. But the father's dissuasive expostulations only persuaded the boy all the more that he would like to be a merchant; and at last, bewildered at his son's self-confidence, he had recourse to the advice of the schoolmaster, who, to his astonishment, was in favour of gratifying the boy's "humour," assigning as a reason that the very fact of his talking about being a merchant showed that there was something in him, for he did not think any other boy in the school knew what the word meant. This advice turned the scales; and with his father's consent young John Brown entered the service of a local firm of merchants named Earl, Horton, and Company, who traded in the staple wares of Sheffield. For the first two years he received no wages, but during the last five years of his apprenticeship he got 6s. a week. He showed a natural aptitude for business, and soon ingratiated himself into the goodwill of his employers. Conscious that his skill and industry were his only avenues to fortune, he sedulously improved every opportunity that could increase his knowledge of the business or strengthen the confidence of his employers. At the end of his apprenticeship his father gave him a suit of new clothes and a sovereign, telling him that for his future success he must rely on his own resources. The indenture of this apprenticeship is still in the possession of Sir John Brown, who regards it as one of the treasures of his palatial residence, Endcliffe Hall.

Important changes in the business of the firm soon opened out what to him was a golden opportunity. In 1836 his employers, who had gone into the steel trade, removed from Orchard Place to Rockingham Street, where they established the Hallamshire Works, and commenced the manufacture of files and table cutlery. In the following year John Brown became of age, and a few months afterwards, to his surprise, Mr. Earl, the senior partner, offered him a share in the business. Being unable to find the capital required, he was obliged to decline this offer; but on further consideration, Mr. Earl offered him the "factoring" part of the business, at the same time promising to aid in finding "the wherewithal" to conduct it. Full of hope and courage, the young merchant preferred to negotiate a loan on his own account, and he succeeded in getting his father and a well-to-do uncle to be security for 500*l.* which a local bank agreed to advance. With this money he bought the business, and thenceforth conducted it with the utmost zeal. Travelling through the country with a horse and gig, he canvassed for his own orders and carried his own samples. His industry soon brought its reward. His business steadily increased; and one extension after another had to be made in order to keep pace with its growth. The gig was succeeded by the four-wheeled sample coach. Instead of retailing the cutlery of other manufacturers, he determined first to make his own goods, and next to make the steel required for those goods. Before entering upon the latter enterprise, however, he asked the consent of his former employers, being unwilling to enter into competition with those to whom he felt indebted for previous favours. The desired consent was readily granted, and in 1844 he commenced the manufacture of steel in small premises in Orchard Street. To the production and application of this metal he then chiefly devoted his attention and resources. His new industry prospered and grew so rapidly that he disposed of his factoring business ; and removing to more suitable premises in Furnival Street, he gave them the name of the Atlas Steel Works, and therein applied himself exclusively to the production of steel, files, and railway springs. The increasing growth of the railway system attracted his attention, and he perceived in it a great field for the consumption of his products. At that time railway carriages or waggons were joined together by a lengthy chain between each of them ; and one reason for arranging them in this primitive way was that there were no buffers in use then. This want was supplied by Sir John Brown. In 1848 he invented and patented the conical spring buffer, which soon proved a great success both commercially and mechanically. He sent the first pair of these buffers ever made to the Taff Vale Railway Company in South Wales ; the second pair went to the Glasgow and South Western Railway ; and the third to the Dublin and Drogheda Railway Company ; while they were first used in England on the London and North Western Railway. All these lines had been constructed about ten years before the conical buffer was invented. The utility of the invention was soon demonstrated, and the demand for the buffers, as well as for other descriptions of railway material, increased rapidly.

An instance of the expedition with which Sir John executed orders occurred about this time. He happened to be in Edinburgh

a few days before the date appointed for the opening of a new
line of railway to Dundee, and calling on the engineer of the
line on Saturday, was informed that everything was ready for the
opening day except a few sets of brake-springs, and that there
was every prospect of the opening ceremony being spoilt through
the default of the contractor in not having these springs ready.
In this strait he asked Sir John if he could supply the springs
by Thursday—five days hence. Sir John pointed out that with
the imperfect means of communication then available he feared
there would not be time to get them. The engineer, evidently
appreciating the energy of the man he was speaking to, said:
"But we must have them." "You shall have them, then,"
rejoined Sir John ; and starting that afternoon by coach for
Berwick, he took the train thence to Newcastle, and thus got
into Sheffield on Sunday forenoon. Arrived at home he sent for
his foreman, and gave orders that his men were to start the
making of these springs first thing on Monday morning, and
that they were to be ready without fail on Monday night. These
orders were executed ; and the springs having been properly
packed, Sir John started with them for their destination. Travelling by rail to Manchester, he there took steamer to Fleetwood,
where it was arranged that a waggon should be ready to carry the
springs to the station from which the mail started for the north.
On reaching the station, however, the railway officials declined to
carry such bulky goods in a mail train. But Sir John was not to
be outdone. He called for the manager, and on telling him the
peculiar circumstances of the case, got a horse-box attached
to the train to carry the springs. In this way he reached
Glasgow on Wednesday afternoon, and there and then delivered
the springs to the engineer, who was so pleased at his expedition
that, in addition to paying the full price and extra expense
involved in the manufacture of the springs, he mentioned the
circumstances under which they were produced to most of the

railway directors present at the opening of the new line ; and for a considerable time afterwards Sir John got most of the orders for the railway material required on Scotch lines.

To supply the increasing demand for these goods he had to extend his means of production, and for this purpose he acquired first one additional workshop and then another, till in 1853 his works were being carried on in four different districts. In the following year an opportunity occurred that enabled him to concentrate his scattered works. In 1854 what were then known as the Queen's Works in Saville Street were offered for sale. The buildings covered an acre of ground, but the site comprised three acres. The works were originally built at a cost of 23,000*l.*, but Sir John Brown bought them for little more than half that amount. On the first of January, 1856, the various detached departments of his business were transferred to this newly acquired property ; and in honour of the event, Sir John, his friends, and workpeople, then numbering about 200, held high festival. The premises were decorated with flags, cannon were fired during the day, the school children of the neighbourhood were feasted, and the workpeople were entertained by their liberal employer at luncheon. After some appropriate remarks from the head of the undertaking, Lady Brown (*née* Mary Schofield) performed " the christening " ceremony by dashing a bottle of wine against the wall, and announcing amid the cheers of the spectators that henceforth the premises were to be named the Atlas Steel and Spring Works. At that time the district around these works was a sylvan retreat, where the wild-flowers blossomed in all the luxuriance of rural repose. How quickly its aspect changed—and such a change ! Its transformation almost realised the description which Milton gives of what in his day was a visionary scene. Ere long it became a place which—

> " Belch'd fire and rolling smoke ; the rest entire
> Shone with a glossy scurf ; undoubted sign

> That in its womb was hid metallic ore,
> The work of sulphur. Thither, winged with speed,
> A numerous brigade hastened: as when bands
> Of pioneers, with spade and pickaxe armed,
> Forerun the royal camp, to trench a field
> Or cast a rampart. Mammon led them on:
> Men also, by his suggestion taught,
> Ransacked the centre, and with their busy hands
> Rifled the bowels of their mother earth
> For (hidden) treasures."

In the course of three years the whole of the three acres of land connected with the Atlas Works were built upon, while the machinery was renewed and enlarged. The year after he entered his new works he embarked in an enterprise which greatly increased not only his own trade, but the trade of South Yorkshire. He determined to try, as an experiment, the manufacture of iron fit for conversion into steel. For that purpose iron of a superior quality was required, and incredible though it may now appear, this country was then dependent upon Sweden and Russia for that quality of iron. Seeing the great demand for iron created by the growth of the railway system, and the high price commanded by the foreign iron used in Sheffield, Sir John Brown, emboldened no doubt by his success in previous ventures, could see no insuperable difficulty in making sufficiently good iron for his own use in his own works. When he first mentioned the matter at a meeting of local manufacturers, they simply laughed at it. But to him mocking laughter only showed the vacant mind. Though the old steel manufacturers pooh-poohed his scheme as a chimera, he satisfied himself of its practicability by proving that the coal, ore, and other materials required in the manufacture of iron, were as cheap in Sheffield as in the foreign countries that then manufactured the iron for the steel trade, and that the steel manufacturers would save by taking the metal in hand in its crude state instead of having first to import it from Sweden and Russia. Accordingly in 1857 he commenced the manufacture of iron, and was so successful that to him is accorded the credit of having inaugurated this new industry in that district. He commenced with six puddling furnaces, a balling furnace, a mill furnace, and two Nasmyth hammers; and the iron produced by these appliances was not only satisfactory in point of quality, but cheaper and more convenient than the foreign iron previously used. Seeing the successful results thus obtained, his competitors in the steel trade, who had formerly derided the scheme, soon came wanting to buy his iron, because they could not get it so cheap from Sweden. In these circumstances the demand for this Yorkshire iron increased rapidly. The six puddling furnaces were soon increased to twelve, which produced about 100 tons of iron per week. Still the demand increased, and further extensions of the works became necessary. His works were bounded on one side by the line of the Midland Railway, and to get more land he had to cross that line. He did so, and on the other side built first one large mill and then another. The first stone of the new works on the north side of the line was laid in June, 1859. In January, 1860, a storm blew down and destroyed half of the roof, measuring 180 feet by 75 feet. This occurred on a Sunday morning, and knowing that Sir John was anxious to get work started in the new premises at the earliest possible moment, a messenger rushed in breathless haste to the old parish church of Sheffield to inform his master of the calamity. When called to the door of the church, Sir John, surprised at seeing one of his workmen with pale face and tears in his eyes, instantly exclaimed: "What is the matter?" "It's all down, sir," replied the workman. "What's down?" "The roof of the new works," continued the workman; "it's blown down." "Then," said Sir John, "go to Harvey and tell him to arrange for putting it up again;" and he coolly returned into the church to hear the sermon. The accident was soon repaired, and in June following the new works were in operation. The iron made at the Atlas Works was then coming into use for other purposes than steel making. Notably boiler and bridge plates were beginning to be made of this Yorkshire iron. Some of the plates used at Charing Cross Bridge were manufactured at the Atlas Works.

It was at this time that Sir John Brown began to produce steel by the Bessemer process. He was one of the five ironmasters, who, immediately after the first announcement of the process agreed to give a large sum for the right to work it in his district; but when it was found that the first expectations were not realised, he, along with the others, abandoned the project. At the beginning of 1859, however, the new works of Henry Bessemer and Company were in operation, and were producing steel at 20*l.* a ton less than the other manufacturers could make it for. When Sir John was designing his new works he intended to produce steel by puddling; but on seeing the Bessemer converter in successful operation next door to his own works, he changed his plan, and asked for a licence to work the Bessemer process. The licence was granted on the terms subsequently charged to all manufacturers of Bessemer steel, namely, a royalty of 1*l.* a ton on steel rails, and 2*l.* a ton on steel for all other purposes. Up to that time railway wheel tyres were sold at 90*l.* a ton; but they could be made of Bessemer steel for 20*l.* or 25*l.* a ton. The tensile strength of the new metal was declared to be 40 tons per square inch, being 15 or 18 tons greater than that of the best Yorkshire iron. As soon as he began to produce this cheap steel, Sir John Brown perceived that it was much preferable to iron for making rails; and accordingly he was the first manufacturer who made rails of Bessemer steel. This he did in 1860. In the following year he informed the Institution of Mechanical Engineers, then holding their meeting in Sheffield, that the new process furnished a pure, homogeneous, hard, and tough material, admirably adapted for the purposes of rail making. In making rails at the Atlas Works in 1861, the ingot of steel was made the exact size, in each case, for a single rail, and in respect of facility of manufacture it had some advantages over the mode of piling by which iron rails were made. He exhibited a number of samples of short steel rails, which had been bent and twisted in an extraordinary manner, without presenting any appearance of fracture. At that time, however, steel rails were much more costly than iron ones; indeed they were then sold at four times the price which they now cost; and hence their slow progress in these early years.[1] Nevertheless the production of Bessemer steel rapidly increased. In 1865 Sir Henry Bessemer told the British Association at Birmingham that steel made by his process was then being used as a substitute for iron to a great and rapidly increasing extent. He added: The jury reports of the International Exhibition of 1851 show that the entire production of steel of all kinds in Sheffield was, at that period, 35,000 tons annually, of which about 18,000 tons were cast steel, equal to 346 tons per week; the few other small cast steel works in the country would probably bring up this quantity to 400 tons per week as the entire production of cast steel in Great Britain. The jury report also states that an ingot of steel, called the " monster " ingot, weighing 24 cwt., was exhibited by Messrs. Turton, and was supposed to be the largest mass of steel ever manufactured in England. Since that date a great change had been made, for the largest Bessemer apparatus erected in Sheffield (in 1865), at the works of Sir John Brown, was capable of producing with ease every four hours a mass of cast steel weighing 24 tons, being 20 times larger than the monster ingot of 1851.

While the new steel works were thus making rapid strides, Sir

---

[1] In later years, when steel rails became so much cheaper, and the demand for them much greater, the Atlas Works ceased to make them, because Sheffield was found to be disadvantageously situated for carrying on a keen competition with works nearer the sea coast that had shipping and other facilities.

John was busily engaged in laying the foundations of another industry, which soon became a large consumer of iron, and with which his name will be for ever associated. He was the pioneer ot armour plate making, not only in Sheffield, but in this country. It is a remarkable fact that, while England is indisputably the first naval power in the world, nearly all the great changes made in the *materiel* of naval warfare have originated in foreign countries. Mr. D. Grant stated, in the House of Commons in 1881, that ironclad armour had originated in France, turret ships in America, torpedoes in Austria, and pebble powder in Germany. In a lecture delivered in May, 1862, " On the iron walls of England," Mr. J. Scott Russell, a great authority on the subject, stated that "the invention of iron armour took place fifty or sixty years ago. I am not prepared to name the first inventor ; but long before we thought of using it in our navy, Mr. R. L. Stevens, a celebrated engineer in New York, the builder of some of the fastest steam vessels on the Hudson, was, I think, the inventor. Certainly Mr. Stevens, between 1845 and 1850, gave me a full account of experiments made in America, partly at his own and partly at the State's expense, and he found that six inches thickness of iron plate armour was sufficient to resist every shot and shell of that day. In 1845 Mr. Stevens proposed to the American Government to construct an iron plated ship, and in 1854 the ship was begun. This ship is in progress, but not yet (1862) finished. Mr. Stevens is, therefore, the inventor of iron armour ; but no doubt the first man who applied it practically for warfare was the Emperor of the French. In 1854 he engaged in the Russian war, and being a great artillerist, he felt deeply what his fleet could not do in the Black Sea, and what we could not do in the Baltic ; and so he put his wise head to work to find out what could be done. In 1854 the Emperor built some floating batteries—four or five ; we simply took his design, and made five or six." Stevens used thin flat plates one over the other ; but Mr. Lloyd, of the Admiralty, on being consulted, expressed an opinion that solid 4½ inch plates would be more effectual than six inches of thickness in a congeries of plates. After the evidence afforded of the success of ironclads in the Black Sea, the Emperor Napoleon determined to make the future fleet of France of iron. Meanwhile, as the Duke of Somerset said, "the House of Commons was in no particular hurry;" and when asked about his own dilatoriness in adopting armour plating, he said, "he delayed until he had consulted the House of Commons about it." In 1856 the Admiralty got the *Trusty* made ready for experiments to test the resistance of ironclad batteries to shot and shell. But after getting her out, the authorities took fright and sent her back again, whereby this country lost two years' start in the construction of its new fleet. In 1855 the design of the *Warrior* was submitted ; but the construction of the first ship of that class was delayed till 1859. In 1858 Sir John Packington first ordered an iron fleet to be made ; but the French had previously commenced the *Gloire*, so that we were three years behind the French. At the close of 1861 we had only the *Warrior* that was fit for service, if it was true, as Sir John Hay, the chairman of the Naval Commission said, that "the man who goes into action in a wooden ship is a fool, and the man who sends him there is a villain."

Although the name of Sir John Brown is now liable to be overlooked in connection with this subject, he took an important, if not a conspicuous part, in the work of transforming our fleet from this obsolete condition to a state of security that no other country has ever approached. The circumstances that led him into this position were of that class which ordinary minds would consider accidental, but which a man of his resources converts into that "tide in the affairs of men, which, taken at its flood, leads on to fortune." He was, in the autumn of 1860, making a tour on the Continent, and happened to be at Toulon when the French vessel already mentioned, the *Gloire*, put into harbour there. This vessel was a curiosity. Originally she was a timber built three-decker, but the French Government had cut down her decks, and covered with armour the portions that were not under water. Some consternation was caused in Government circles in England by the announcement that this "ironclad" had been put in commission; and our Government, not having at that time a single ironclad, determined to convert ten large wooden men-of-war into the shape adapted for armour plating. Sir John Brown perceived that in the production of iron plates for armour clads there was a new field for his enterprise and skill. He therefore asked permission to go on board the *Gloire* for the purpose of examining the armour plating, but this was refused. Determined to succeed, he made as minute an examination of the exterior of the vessel as he could from the nearest point of view; and from this inspection and inquiries he ascertained that the armour plates were 5 ft. long by 2 ft. wide, were $4\frac{1}{2}$ in. thick, and were made by hammering. He thought he could make thicker, larger, and tougher plates by rolling the iron instead of hammering it; and he returned to Sheffield with the intention of putting his ideas to a practical test. He erected a rolling mill, selected workmen, and personally directed the operation to a successful issue. The way in which he manufactured a five-ton armour plate was thus described by himself at a meeting of the Institution of Mechanical Engineers in Sheffield :—Several bars of iron were rolled 12 in. broad by 1 in. thick, and were cut 30 in. long. Five of these bars were piled and rolled down to a rough slab. Five other bars were also rolled down to another rough slab; and these two slabs were then welded and rolled down to a plate $1\frac{1}{2}$ in. thick, which was sheared to 4 ft. square. Four plates like that one were then piled and rolled down to one plate measuring 8 ft. by 4, and $2\frac{1}{2}$ in. thick. Lastly four of these were piled and rolled to form the final and entire plate. There were thus welded together 160 thicknesses of plate, each of which was originally 1 in. thick, to form one plate 4½ in. thick, being a reduction of thirty-five times in thickness; and in the operation from 3,500 to 4,000 square feet of surface had to be perfectly welded by the process of rolling. It was not surprising, he added, that even with the greatest care blisters and imperfect welding should occur and render the plate defective. This was the chief difficulty to be overcome, and it increased with the magnitude and weight of the plate, the final operation of welding the four plates, measuring 8 ft. by 4 ft., and 2½ in. thick, being a very critical one. The intensity of the heat thrown off was almost unendurable, and the loss of a few moments in the conveyance of the pile from the furnace to the rolls would be fatal to success.

No sooner had Sir John Brown made a fair start in the manufacture of armour plates by this process than some formidable competitors entered the field and openly contested his claims to superiority. The first orders of the Government were divided among different manufacturers, Sir John getting a share, at prices ranging from 37*l.* to 45*l.* a ton. In September, 1862, the question of superiority was put to a decisive test. Experiments were made at Portsmouth with four plates forged in the Government dockyards and with one manufactured at the Atlas Steel Works. The latter was selected from a heap of plates which had been made for the *Royal Sovereign* shield ship, and it weighed 94 cwt. 3 qr. These plates were secured to the side of the *Alfred* target ship, and were fired at from the 95 cwt. gun of the *Stork* gunboat. Solid 68 lb. shot were thrown with the ordinary 16 lbs. of powder. According to a contemporary account, unusual interest attached to the trial, because the plates from the Government yards had been manufactured purposely to test the cost of production in comparison with the price paid to contractors (which some detractors had represented as exorbitant), and also to decide upon the respective merits of puddled and scrap iron as the material for plating. The plates made at the Government yards soon broke up, while the plates produced by Sir John Brown stood a severer ordeal than any plates had ever been subjected to before. Four shots sufficed to destroy the Government plates, while Sir John Brown's appeared invulnerable after receiving nine shots. In subsequent tests he likewise carried off all the honours.

While some powerful competitors advocated hammering as the best means of making armour plates, Sir John continually advocated the rolling system, and demonstrated that it possessed many great merits. Nearly all the experimental plates required by the special commisson on iron plates appointed by the Government were made at the Atlas Works. Many costly experimental plates were supplied from these works free of charge. At the close of the Exhibition of 1862 Sir John was awarded the gold medal for excellence in armour plating : and it may be here added that in 1867 he received from the French jurors the sole gold medal for British armour plates. Members of the English Government also acknowledged his pre-eminence as a maker of armour plates. In August, 1862, Lord Palmerston, then Prime Minister, went to Sheffield on a special visit to Sir John Brown at his residence, Shirle Hall, and took a lively interest in the various processes carried on at the Atlas Works, where he saw a plate rolled that weighed six tons. Speaking on the navy estimates in the House of Commons in February follow-ing, Lord Clarence Paget said : " While upon armour plates let me pay honour where honour is due. We can get good plates, both hammered and rolled, but we find that the rolled are more uniform ; and Sir John Brown, a gentleman distinguished by great zeal, and conducting important works at Sheffield, has been most successful in producing these plates. "

Up to 1863 some of the naval heads of the Government

thought it next to impossible to produce armour plates more than 4½ in. thick; but Sir John Brown, who saw that the increased power of the artillery coming into use would soon render such thin plates useless for defensive purposes, was then erecting a new rolling mill designed to produce larger plates. In reference to the strength of the armour plates that were first made, Sir Joseph Whitworth relates the following : " I remember telling the Duke of Somerset when I penetrated the *Trusty*—I think I sent four shots through the *Trusty*—that I had no doubt from what I saw that I should be able to make shell go through. That created immense surprise. He had no idea that such a thing could ever be accomplished as sending a shell through armour plating. My steel projectile in 1858 was the first to do it. At 450 yards the first shot that was fired went through it."

In these circumstances Sir John Brown offered the Government to roll three plates, 5, 7, and 8 in. thick respectively ; and if they failed to resist the shot that penetrated the 4½ in. plates, he would bear the cost of the experiments. On his invitation, the Lords of the Admiralty and other noblemen attended to witness the opening of the new mill, and to see the " monster " plates rolled. This took place on the 9th of April, 1863, when there were present the Duke of Somerset, Lord Clarence Paget, the Marquis of Ripon, the Duke of Devonshire, Earl Fitzwilliam, Lord Wharncliffe, and other distinguished men. They saw several plates rolled exceeding 4½ in. in thickness, and the operations concluded with the rolling of a plate 12 in. thick, measuring 15 ft. by 20 ft., and a plate 5 in. thick, measuring 40 ft. by 4 ft. Addressing the workmen, Sir John said : " We are all proud of your exploits; you are all worthy of the name of Englishmen. His Grace the Duke of Somerset wishes me to express his admiration of what you have done."

At the banquet which followed the Duke said that Sir Joseph Whitworth maintained that whatever plates were made he would fire through them. "I always encourage him," continued his Grace, "to give us the most irresistible artillery he can, because we wish to have irresistible ships, and, on the other hand, armed with irresistible artillery. These are the difficulties in which we are at present placed, and I must now say that what I have seen to-day gives me hopes on the one side, that as to the protection of our ships we are now in a fair way of meeting the difficulty." He also praised the intelligence, good temper, and kindly feeling of the workmen in the Atlas Works, saying he was convinced that the men felt they were well treated, and that the head of the establishment managed them with great judgment and kindness—which was the only way in which such great works could be carried on. In proposing prosperity to the new rolling mill, he said it would be "in the future one of the most wonderful pieces of machinery ever made in this country."

Punch published a characteristic account of this event, which is still worth reading:—

"Now," said Mr. Punch, "let the ceremonies proceed. Somerset, my boy, do you think you understand anything about the process?"

"Well, yes," said the first Lord of the Admiralty, "I think I do. You see, they make it hot, and then——"

"Make what hot? Brandy and water? That reminds me that I should like a little, for I am far from well."

"I mean the iron," said the Duke, when Mr. Punch had finished the liquid that was tendered to him as he spoke.

"Well, why didn't you say the iron? Didn't you like to speak ironically?"

It is well that Mr. Brown had built his works strongly, for a shout like that which followed would have brought down any light erection.

"Well," said the Duke, "they take it out of the furnace and roll it between these rollers, and that is all."

"Not quite," said the Mayor,[1] with a quiet look at Mr. Punch ; "but his Grace is not altogether an unintelligent observer. Here comes a plate."

The brawny giants suddenly drew open the door of a vast furnace, and you had an idea that a large piece of blazing fire had got in there by accident and it was about as possible to look in the face of the fire as of Phœbus. Then, tugged forth by the giants, out came a large slab of red-hot metal, just the thing for a dining-table in Pandemonium, and it was received upon a mighty iron truck, and hurried along to the jaws of the rolling machine. As it was drawn fiercely into the mill a volcano broke out, and the air was filled with a shower of fire-spangles of the largest construction, and eminently calculated to make holes in your garments. The monster slab was so mercilessly taken in hand by the mighty wheels, and was hurled backwards and forwards, under terrific pressure, and so squeezed and rolled and consolidated, that when at length it was flung, exhausted as it were, upon the iron floor beyond, Mr. Punch was reminded of the way in which he has dealt with, improved, and educated the public mind for the last twenty years.

"And that's the way I propose to defend the British navy," said the Duke of Somerset, looking as if he had done it all.

"Mr. Mayor," said Mr. Punch, "it makes me thirsty to hear these aristocratic muffs going on in this manner. I hear you have spent 100,000*l.* in this single part of your works in six months, and that you are going to build largely in addition. Sir, I suppose that we, the nation, shall have to pay you a trifle for what you manufacture."

Mr. Brown smiled, as if he thought that just possible.

"Sir," continued Mr. Punch, "I rejoice thereat. I don't care what these things cost. I consider them the cheap defence of nations, at least of our nation, which is the only one I care a red

[1] Sir John Brown was then Mayor of Sheffield.

cent about. These things will make war as nearly impossible as anything in this mad world can be; and therefore, Mr. Brown, I hope you will go on making them until further notice."

The large plates which the Admiralty tested were found so satisfactory that orders were given that they were to be paid for; and henceforth Sir John Brown became the largest maker of iron plates for the Government. His fame as a manufacturer of these plates soon spread abroad. At one time it was announced that plates were being produced in France that could successfully compete with those of British manufacturers in the markets of Europe, and no little consternation was caused by the successful results obtained at the trial of some French plates at Portsmouth. How these successful French plates were produced was not known, but the failures sustained by French manufacturers afterwards were so rapid, that they were driven out of the market nearly as quickly as they entered it. They were officially condemned in Russia, Holland, Sweden, England, France, and Denmark. In the latter country testing experiments took place in the last week of 1863 upon five plates, sent from Lyons, Glasgow, the Thames, and Sheffield. The one made by Sir John Brown showed greater powers of resistance than any of the others; and he was accordingly awarded the first order of merit. He did not, however, immediately supply these foreign customers. One foreign Government after another applied for his plates, but he refused to supply them while he was busy executing orders for his own Government. In 1867 it was reported that three-fourths of the ironclads of the British navy were clothed with armour plates made at his works.

Being then the acknowledged leader in the manufacture of armour plates, the demand for them and for his other products was such as required almost continuous extensions to be made to the Atlas Works. In the development of his plate rolling mills he expended 200,000*l.* In 1857 his works covered a single acre; in 1867 they covered 21 acres. The buildings which closely covered this large area were all designed by himself without the assistance of an architect, and they were filled with machinery used in the production of plates, ordnance, forgings, railway bars, steel springs, rails, tyres, axles, &c. Much of the machinery was designed by himself, and all of it was made under his supervision. A single incident will give an idea of the difficulties he had to surmount in furnishing his workshops with the requisite machinery, even of an ordinary kind. When he commenced the manufacture of armour plates he was unable to find in any of the leading machine shops in the country planing and slotting machines large enough to finish such immense masses of iron. A leading machine maker in Glasgow was astonished to hear Sir John say that his largest productions were too weak, and assured him that nothing he could do would break them. "Ah!" said Sir John, "you don't know what I want them for; nothing made at present may break them, but I want them made stronger." Calling for the drawings of the largest machines, he marked with a pencil the parts that were to be made stouter and stronger, so as to stand a greater strain than they had ever been subjected to before, and then handing the design to the manufacturer, ordered machines to be made of these enlarged dimensions, agreeing to pay so much a ton for them. By his order steel shafts were put into them instead of wrought iron, and when finished these machines were found to be twice as strong as any hitherto produced. Though he was continually devising and introducing labour-saving machinery, the hands employed at the Atlas Works increased as rapidly as the works. In 1857 they numbered 200; in 1867, 4,000. In the first year of his business he turned over about 3,000*l.*; and in the last named year it was nearly 1,000,000*l.* In connection with his works he raised two corps of volunteers, and equipped them at his own cost. In 1864 the works were registered as a limited liability concern, with a capital of 1,000,000*l.* Sir John became chairman of the company, and Mr. J. D. Ellis and Mr. W. Bragge, whom he had previously taken into partnership, became managing directors.

After the retirement of Sir John from the active duties of management, the Atlas Works continued to display a degree of enterprise and skill that entitled them to a foremost place in the history of the steel trade. It was at these works that chrome steel was first made in England. For many years it had been known that a mixture of chromium and iron could be made to produce steel of great hardness and strength ; but it was not till 1871 that it was brought into practical use. Mr. Julius Bauer patented in America a way of producing chrome steel in crucibles, and the metal produced by his process was described as having extraordinary properties. It was said that the new steel, being an alloy, was capable of being graded for any special purpose, that it could be made so hard that it could not be softened, and so soft that it could not be hardened ; that it had a tensile strength far exceeding that of any other kind of steel, and that one grade of it, called adamantine, when forged into a tool and allowed to cool gradually, was too hard to be worked with a file. Chrome steel was used for those parts of the St. Louis Bridge, U.S., in which great strength was required. In 1875 Sir John Brown and Co. took up the manufacture of this steel, about which little or nothing was then known in this country, and they claimed for it a remarkable degree of strength, malleability, and freedom from corrosion. This steel, however, never came largely into use in this country, though some makers of edge tools have a decided preference for it.

In the manufacture of armour plates the Atlas Works continued in the van of progress, notwithstanding the skill and enterprise of able competitors. In later years this department of the works was further improved and adapted for the most recent requirements of the trade. From the memorable time when Sir John

Brown demonstrated that he could roll plates of greater thickness
than 4½ inches, increased thickness was one of the features of our
ironclads built before 1876. When the *Inflexible* was commenced
in 1874, it was intended to arm her with 24 inches of iron plating,
but before she was ready to receive her iron walls an important
change in the material and manufacture of armour plates took
place. At Spezzia in October, 1876, the 100 ton gun completely
perforated and smashed 22 inches of iron plates and their back-
ings. This appeared a fatal blow to iron. The French manufacturers thereafter directed attention to the power of steel plates,
which the Italian Government thought superior to iron ones. At
the same time Sir Joseph Whitworth obtained remarkable results
with plates of steel, while the Sheffield manufacturers declared in
favour of iron plates with steel faces. In 1877 Mr. J. D. Ellis, of
the Atlas Steel Works, took out a patent for the manufacture of
these steel-faced plates; and at the Paris Exhibition of 1878 some
of " Brown's compound plates "—half iron and half steel—were
exhibited for the purpose of showing their power of resisting a
more severe trial than iron ever withstood. The process of
manufacture was this : an armour plate, made in the usual way,
had affixed to it a frame made of iron bars, so as to inclose on
the plate a quantity of steel equal to about half the thickness of
the iron plate. The plate thus "framed" was placed in a heating
furnace till it reached a welding heat, after which it was withdrawn, and molten Bessemer steel was poured onto it. This is
an interesting process to see. At a given signal a ladle, containing about ten tons of steel, is carried along by a powerful crane,
and being suspended over the plate, and the plug withdrawn, the
liquid steel flows in a stream into the box-like frame. Sometimes
the stream of metal is interrupted to prevent it settling down, and
to secure an equal distribution all over the plate. Sometimes the
metal bursts forth at the side, and falls in a shower of star-shaped
sparks, not unlike the beautiful phenomenon of the Bessemer

process. When there is more metal in the ladle than is needed for the plate, the surplus is run into moulds to form ingots for other purposes. Meanwhile the compound plate is allowed to cool, and being afterwards again re-heated, is rolled in the usual manner. A plate which, after receiving its steel facing, is $16\frac{1}{2}$ inches thick and weighs 20 tons in the rough, is reduced before being finished to 9 inches in thickness and 14 tons in weight. The plates thus produced have been subjected to all sorts of trials, and found equal, if not superior, to all competitors. The projectile of the 100 ton gun only penetrated 8 inches into a 19 inch steel-faced plate, whereas, if it had been made of iron, it would certainly have been perforated.

A sample armour plate, manufactured at the Atlas Works, for the Russian Government, was tested in 1883 on board the *Nettle* at Portsmouth. It was a specimen of the belt protection of a Russian cruiser in course of construction, and measured 8 feet by 7 feet, the thickness tapering from 6 inches to 4 inches below the water-line, and the steel face forming one-third of the thickness. Three rounds were fired from the 7 inch gun, with charges of 14 lb., and projectiles of 114 lb. The test was highly successful. Nothing was produced except hair cracks, of which the most important was only about one-tenth of an inch wide.

Reporting in 1883 on the manufactures and exports of Sheffield, Mr. C. B. Webster, the American consul in that town, stated that the firm of Sir J. Brown and Co. "use at their Sheffield works 160 steam boilers, with between 11,000 and 12,000 h.-p. They have the largest planing machines—used for finishing armour plates—in the world. Their weekly pay roll amounts to 7,000*l.* They consume a quarter of a million tons of coal and coke annually. I saw, at their works, a day or two since, an armour plate 18 inches thick, weighing from 20 to 23 tons, made for the English Admiralty; also several heavy beam plates for shipment to the United States. The largest plate recently rolled is for the

Italian Government. It is 19 inches thick, and weighs, when finished, over 32 tons. These plates are a combination of iron and steel, the patent of Mr. Ellis, the chairman of the company. The total weight of armour for the large Italian vessel *Italia* will be about 18,000 tons. In addition, she will require iron deck-plates from 3 inches to 4 inches thick, weighing 800 or 1,000 tons. The process of handling and bending these immense armour plates to fit every part of the ship is an interesting one, and their edges are so nicely planed, that when placed in position they fit each other with the utmost exactness. This firm illustrates the value of a name. When Sir John Brown, its founder, left the business to his successors, he was paid 200,000*l.* for goodwill and for the privilege of retaining the well-known title."

Not only did these works increase the industrial resources of Sheffield, but they called forth the emulation and the energies of worthy competitors in the same direction. The great progress made in the staple industries of the town, in which Sir John was a pioneer, is reflected in the increase of its population. When the present Atlas Works were started in 1857 the population of Sheffield was about 133,000; in 1881 it was 284,000. Apart from the effect of his industrial enterprise, he has personally rendered invaluable services to the town in which his life has been spent. In all works of charity, and in all movements calculated to promote its advancement, he has taken a foremost part. His life has been marked by great liberality and a desire as far as possible to benefit those around him; and in recognition of his public spirit, his townsmen have conferred upon him all their local honours. In the consecutive years 1862-63 he was mayor of the borough, and after that he remained an alderman. There are two public bodies in Sheffield which hold and administer property for the benefit and improvement of the town—the church burgesses and the town trustees; and he was made a member of both these bodies. He was also made a magistrate for the borough, as well as for the North West Riding, of which, too, he was appointed a deputy lieutenant. Twice he held the office of master cutler; and his tenure of that office was signalised by the completion of the Cutlers' Hall. The honour of knighthood was conferred on him in 1867. When, in 1871, the Sheffield School Board was constituted, he was elected chairman; and ten years afterwards a bust of him was placed in the board room of that body to commemorate his services. In works of charity he was likewise a man of "light and leading." His name appeared as a patron or officer in many local societies. His liberal support of every movement likely to benefit the town was more conspicuously shown during the years he acted as mayor. It was stated afterwards that during those two years he spent 6,000*l.* for the benefit of the public. Through his efforts a large sum of money was raised in Sheffield in aid of the Lancashire distress fund; and when, immediately after his retirement from the civic chair, the memorable inundation occurred through the bursting of a reservoir, causing a serious loss of life, he took a foremost part in alleviating the sufferings which that calamity entailed. He visited the most urgent cases of distress, and supplied their immediate wants from his own purse. In the district in which most of the large works are situated, he built a church and schools in 1867-8 at a cost of 12,200*l.* Its ecclesiastical name is "All Saints;" its local name is "John Brown's Church." In opening this church on February 5th, 1869, the Archbishop of York said, in reference to the donor, "I feel persuaded, from many conversations and from what I know, that the feeling uppermost in his mind was not to raise a grand temple, which, seen from afar by men, would be an ornament to the town and a monument of his own liberality; I feel sure it was his great anxiety to do what he could towards saving the souls of those who work for him." Although himself a Churchman, it is said that there is no religious denomination in Sheffield that has not had reason to appreciate his generosity.

MR. SIDNEY GILCHRIST THOMAS.

CHAPTER X.

" The invention all admired, and each how he
  To be the inventor missed; so easy it seemed
  Once found, which, yet unfound, most would have thought
  Impossible."—MILTON.

RARELY has an inventor attained world-wide celebrity so quickly as Mr. Sidney Gilchrist Thomas. He solved a problem which had baffled the greatest metallurgists since the invention of the Bessemer converter. The trouble which the presence of phosphorus in iron occasioned to Sir Henry Bessemer has already been recorded ; but other metallurgists, as well as he, had tried to effect its elimination. Among this array of eminent but unsuccessful experimentalists were Karston, Tanoyer, Wall, Winkler, Fleury, Guest, Evans, Englehart, Knowles, Heaton, Hargreaves, Fuchs, Crawshay, Fissier, Warner, Drown, Troost, Daelen, Rochussen, and others. To these has to be added one of the greatest and one of the oldest of contemporary metallurgists—Mr. Isaac Lowthian Bell. He had for years been regarded as the high priest of British metallurgy ; and the fact that he did not solve the problem, after very elaborate investigations, was by many considered a proof that it was insoluble. It was believed that no one knew more than he did about the manufacture of iron in general, and the baneful nature of phosphorus in particular. Besides being the chief proprietor of the largest works in England that produced pig iron only, his observations on the phenomena of the blast furnace extended over a period of nearly forty years. In 1870-2 he communicated to the world the fruits of his studies in an elaborate work entitled *The Chemical Phenomena of Iron Smelting*, which recorded the results of about 1,000 experiments, besides collating the innumerable data and experiments of other metallurgists past and present, British and foreign. The concluding words of that work contained some interesting references to the importance of eliminating phosphorus from iron. He said : " The very cheapness of iron has been the means of introducing its use in a thousand ways, to which high price would have shut the door, and when a better article for higher class work was required, it was easier and less expensive to go at once to better class iron, than engage in costly experiments for the purpose of freeing the cheaper article of its imperfections.

" Such was the state of things a few years ago, when the cost of producing a ton of pig iron, free from phosphorus, probably did not exceed by 10s. that of Cleveland, with its 1 or 1½ per cent. of this element.

" The introduction of Bessemer steel for railway bars, the necessity of constructing our locomotives and iron steamers of great strength, combined with great lightness, have changed all this. Steel is now a form in which iron will be greatly sought after, and in such anxious request is pig iron, suitable for the manufacture of this material, that it has run up rapidly from about 60s. to nearly 6l. per ton, being nearly double that of pig iron obtained from Cleveland stone.

" The limit to the production of Bessemer pig is want of ores free from phosphorus. The hematites of this country, under the sudden demand, have doubled in price, and speculators of all kinds are rushing off to Spain, where tracts of land, conceded without any payment a few months ago by the Government of that country, are said now to be worth large premiums; at least, such is the impression left on the mind by a perusal of the published prospectuses of the day.

"This may be correct, and so firm may be the grip that phosphorus holds on iron, that breaking up the bonds that bind them together may defy the skill of our most scientific men ; but it may be well to remember that the yearly make of iron from Cleveland stone alone contains about 30,000 tons of phosphorus, worth for agricultural purposes, were it in manure as phosphoric acid, above a quarter of a million, and that the money value difference between Cleveland and hematite iron is not short of four millions sterling, chiefly due to the presence of this 250,000*l.* worth of phosphorus.

"The Pattinson process does not leave one part of silver in 100,000 of lead ; the Bessemer converter robs iron of almost every contamination except phosphorus, but nine-tenths of this ingredient is expelled by the puddling furnace. It may be difficult, but let it not be supposed there would be any surprise excited in the minds of chemists, if a simple and inexpensive process for separating iron and phosphorus were made known to-morrow, so that only one of the latter should be found in 5,000 of the former ; and now that there is such a margin to stimulate exertion, we may be sure the minds of properly qualified persons will be directed towards the solution of a question of such national importance."

No one gave more evidence of diligent application to the consideration of this problem than the author of these words. In 1877 he read two papers before the members of the Iron and Steel Institute on the conditions which influenced the separation of carbon, silicon, sulphur, and phosphorus from iron as they exist in the pig. He showed as the result of numerous experiments that carbon and silicon were expelled at moderately high, as well as at the more intense, temperatures commanded by the different furnaces and apparatus employed in the manufacture of iron and steel ; and that the agent which effected the removal of these two substances was oxygen. In some, he said, this is probably performed by direct action, as in the Bessemer converter, while in others the oxidation of the carbon and silicon is chiefly produced by the fluid cinder, in which oxide of iron is the active body. He was of opinion that the separation of sulphur took place in a similar manner ; but phosphorus appeared to be influenced by a different condition of things. Oxygen in its free state was almost entirely inert as regarded phosphorus at the intense temperature of the Bessemer process. But he found that when melted pig iron was exposed to the action of oxygen at lower temperatures, phosphorus was rapidly removed. He therefore held that the order of affinity between iron and phosphorus, by difference of heat alone, was inverted. This was ascertained by a variety of experiments, and he gave details of several which showed that phosphorus tended to disappear at a low temperature and to return at a high temperature.

At the annual meeting of the Iron and Steel Institute in 1878, the first paper read was one by Mr. Isaac Lowthian Bell, " On the separation of phosphorus from pig iron." In this communication he explained the mechanical apparatus in which he had so applied the fused oxides of iron to liquid iron, at a comparatively low temperature, as to remove 96 per cent. of the phosphorus ; but he left the commercial value of the process untouched. As soon as the paper was read, Prof. Williamson expressed the obligation the members were under to Mr. Bell for his valuable information. Mr. Snelus next rose and stated that six years previously he had taken out a patent for using lime for the lining of steel-melting and other furnaces ; and although he had refrained from saying anything about it at these meetings, he was so impressed with " the value of the essence of the thing," that he had been trying during these six years to devise some mechanical means of reducing it to practice. In the middle of the discussion a young member, apparently the youngest man in the meeting, modestly stated in three sentences that he had succeeded in effecting the almost complete removal of phosphorus in the Bessemer process; and that some hundreds of analyses made by Mr. Gilchrist showed a reduction of from 20 to 99·9 per cent. of phosphorus. The meeting did not laugh at this youthful *Eureka*, nor did it congratulate the young man on his achievement. Much less did it inquire about his method of elimination; it simply took no notice of his undemonstrative announcement. The young man, whose name probably very few of those present had ever heard before, was Sidney Gilchrist Thomas, and the Thomas-Gilchrist process was then announced for the first time. Future historians will probably date from that announcement a revolution in the steel trade of the world; and its author's name at present stands foremost among the successful inventors of steel.

Born in 1850, Sidney Gilchrist Thomas was educated at Dulwich College, near London, where he received a purely classical training with the view of studying medicine. On finishing his studies there he was about to proceed to London University, with the view of graduating in arts, but his father dying at that critical period, he determined not to proceed further in his academic career. He resolved to make his way in the world by a shorter cut, and accordingly he became for a short time a teacher in a private school. At the age of seventeen he entered the Civil Service, but not with the intention of remaining in it. Though he continued in the service till 1879, he turned these years to good account, with the view of attaining quicker and more valuable distinction than the service could give. He always showed a predilection for science, and impelled by the love of it, he worked assiduously in his leisure hours to master the elements of chemistry. His intention was to go into metallurgy, and in order to pursue his chemical studies he had a small laboratory fitted up for his own use. He also studied occasionally in the evenings at Mr. Vacher's laboratory in London. Having fixed upon metallurgy as the special branch of science which he should pursue, he took care to qualify himself for passing the examinations of the School of Mines. The curriculum of that school extends over three years, but he did not attend the lectures during that period. Dr. Percy was then the lecturer, but Mr. Thomas's other engagements did not admit of his attendance. Nevertheless he passed all his examinations except that of metallurgy, which is only open to those who attend the whole course of lectures. He generally spent his holidays in visiting iron works in this country and on the continent, in order to gain a better insight into the different operations and methods in use for the smelting and refining of metals. In his laboratory studies he made it a practice to select three or four problems in connection with things unsolved, so as to group facts around them with the view of seeing how far he could obtain a solution of them. It was in this way that he took up dephosphorisation, at which he worked at intervals for seven years. In order to master the known conditions of the problem, he first collected all the analytical and technical data available on the subject. He soon came to the conclusion that the best practical way to eliminate the phosphorus was to obtain a very strong base, which should be added to the Bessemer process to enable the oxydised phosphorus to unite with it and thus be carried off in the slag. The term *base* is used by chemists to signify a compound which will chemically combine with an acid ; and the phosphorus, when oxidised in the Bessemer converter, is technically called phosphoric acid. In other words, the base and the acid have a " liking " for each other, and the one thus combining with the other, they could be expelled together. For this purpose it seemed clear to him that a basic lining must be used. Having arrived at these *primâ facie* conclusions as to the best method of procedure, he entered upon a series of experiments for the purpose of investigating the nature and duration of different sorts of linings, and soon became convinced that the requisite material must be either lime or magnesia. He then entered upon some experiments with a Bessemer converter on a miniature scale in London, but finding it very difficult to obtain the pressure of blast necessary to carry such experiments further, he wrote to his cousin, Mr. P. C. Gilchrist, who was then chemist at Cwm-Avon in South Wales, laying the condition of the experiments before him, and asking his co-operation with the view of experimenting with a converter on a greater scale. Nothing was done just then, but soon afterwards Mr. Gilchrist went to Blaenavon as analytical chemist, and there he made arrangements for further experiments in the direction indicated by Mr. Thomas. Accordingly these two young men—the one aged twenty-six and the other twenty-five years—carried on a series of experiments for eighteen months with crucibles lined with lime, oxide of iron, magnesia, and other substances; and after a long delay they got a miniature converter, which, though it only held eight pounds instead of eight tons, sufficed, when supplied with blast from the furnace, for their experimental purposes. Another series of experiments was then begun, and with prospects of success. About midsummer in 1877 they began to obtain satisfactory results, which proved that Mr. Thomas's theory was right. They effected partial dephosphorisation in many instances with North-ampton pig by lining the converter with bricks of limestone and with silicate of soda, which lasted fairly well; but owing to some defect in the apparatus they were not able to get a cast fluid so as to finish the operation. Early in the autumn of the same year they got a number of casts of eight pounds each which were completely dephosphorised, and which on analysis were found to be excellent steel. These results they communicated to Mr. Martin, of the Blaenavon Works, who offered them further facilities for carrying on their experiments. Till then they had been making them, for the sake of secrecy, in an old smithy shed; and Mr. Martin undertook, on behalf of the Blaenavon Company, to build a small kiln for making their bricks, and to let them have the use of a small converter—small compared with the usual size, but large compared with the miniatur one they had been using. Some time before that they had been experimenting on the effect of a very high temperature in producing a very hard and compact structure in the limestone, and for that purpose they used a Fletcher injection furnace. They found that by exposing the magnesia and limestone to a very intense white heat for a considerable period, it became shrunk lime, and as hard as flint. They at once saw that this material was likely to withstand the intense heat in the converter, and to contain within its compact physical structure the base necessary to eliminate the phosphorus. Believing that they had at last obtained the means of practically solving the problem, they tried to manufacture these bricks in the small kiln built for the purpose at Blaenavon ; but they found that the fireplace was too small, and the heat insufficient to thoroughly burn them. In the winter of 1877 they had two converters at their service—a fixed one holding four hundredweight, and a tipping one holding ten hundredweight. The rights of the inventors were now secured by patent ; and the results already obtained having been communicated to the Dowlais Works in February or March, 1878, arrangements were made for the use of a large converter there. Some experiments were made in a 5-ton converter at Dowlais, but they were unsuccessful ; dephosphorisation took place, but they did not get more than five casts. The inventors elected not to proceed further there, because the floor of the kiln prevented them getting a proper supply of lime bricks for lining the converter. But, with the vigorous co-operation of Mr. Martin, experiments were resumed at Blaenavon, and continued till September, when an account of the results of their labours was prepared for the Paris meeting of the Iron and Steel Institute. During an excursion of the members of that institute from Paris to Creusot, Mr. Thomas mentioned to Mr. E. W. Richards, the manager of Bolckow, Vaughan, and Co.'s works in Cleveland, the state in which their experiments stood, and their desire to continue them on a larger scale.

Mr. Richards, who is not only a practical ironmaster of large experience, but a gentleman much esteemed for the liberality and kindliness with which he is ever ready to lend a helping hand to a good cause, immediately took an interest in Mr Thomas's representations; and his own account of his honourable connection with the working of the process forms one of the most interesting parts of its history. In his presidential address to the Cleveland Institution of Engineers on the 15th of November, 1880, Mr. Richards said: "Messrs. Thomas and Gilchrist prepared a paper giving very fully the results of their experiments, with analyses. It was intended to be read at the autumn meeting of the Iron and Steel Institute in Paris in 1878; but so little importance was attached to it, and so little was it believed in, that the paper was scarcely noticed, and it was left unread till the spring meeting in London in 1879. Mr. Sidney Thomas first drew my particular attention to the subject at Creusot, and we had a meeting a few days later in Paris to discuss it, when I resolved to take the matter up, provided I received the consent of my directors. That consent was given, and on the 2nd October, 1878, accompanied by Mr. Stead of Middlesbrough, I went with Mr. Thomas to Blaenavon. Arrived there, Mr. Gilchrist and Mr. Martin showed us three casts in a miniature cupola, and I saw sufficient to convince me that iron could be dephosphorised at high temperatures. I also visited the Dowlais Works, where Mr. Menelaus informed me that the experiments in the large converters had failed, owing to the lining being washed out. We very quickly erected a pair of 30-cwt. converters at Middlesbrough, but were unable for a long time to try the process, owing to the difficulties experienced in making basic bricks for lining the converters and making the basic bottom. The difficulties arose principally from the enormous shrinkage of the magnesian limestone when being burnt in a kiln with an up-draught, and of the failure of the ordinary bricks of the kiln to withstand the very high temperature necessary for efficient burning. The difficulties were, however, one by one surmounted, and at last we lined up the converters with basic bricks ; then, after much labour, many failures, disappointments, and encouragements, we were able to show some of the leading gentlemen of Middlesbrough two successful operations on Friday, April 4th, 1879. The news of this success spread rapidly far and wide, and Middlesbrough was soon besieged by the combined forces of Belgium, France, Prussia, Austria, and America. We then lined up one of the 6-ton converters at Eston, and had fair success. The next meeting of the Iron and Steel Institute in London, under the presidency of Mr. Edward Williams, was perhaps the most interesting and brilliant ever held by that institute. Messrs. Thomas and Gilchrist's paper was read, and the explanations and discussions by other members of the institute were listened to with marked attention.

" Directly the meeting was over, Middlesbrough was again besieged by a large array of Continental metallurgists, and a few hundredweights of samples of basic bricks, molten metal used, and steel produced were taken away for searching analysis at home. Our Continental friends were of an inquisitive turn of mind, and, like many other practical men who saw the process in operation, only believed in what they saw with their own eyes, and felt with their own hands. And they were not quite sure even then, and some are not quite sure even now. We gave them samples of the metal out of the very nose of the converter. Our method of working at that time was to charge the additions of oxide of iron and lime at the same time into the converter, and pour the molten metal upon them. The quantity of additions varied from 15 to 25 per cent. of the metal charged, according to the amount of silicon in the pig iron used. We soon found that the oxide of iron was unnecessary ; besides, it cooled the bath of metal ; and we after-

wards used lime additions only. After about three minutes' after-
blow, a sample of metal was taken from the converter, quickly
flattened down under a steam-hammer, and cooled in water. The
fracture gave clear indications of the malleability of the iron.
When the bath was sufficiently dephosphorised to give a soft,
ductile metal, the spiegel was added. Other firms have taken up
the manufacture of steel on the basic system, notably the Hôerde
Company in Westphalia, and Messrs. Brown, Bayley, and Dixon
in Sheffield. Very interesting papers on the subject have been
read by Messrs. Pink and Massenez and Messrs. Holland and
Cooper. On Monday, the 23rd of August, (1880), I visited the
Hôerde Works with a few friends, and saw two successful casts in
a small converter. Imitating the good example set me, and
having good friends in Messrs. Pink and Massenez, I took a
sample of the re-melted pig as it was running from a cupola to
the converter, and a sample of dephosphorised metal and of the
steel. Our chemist's (Mr. Cook's) analysis of the re-melted pig
is :—combined carbon, 2·75 ; manganese, ·50 ; silicon, ·9 ;
sulphur, ·31 ; phosphorus, 1·51. This analysis agrees with that
given by Mr. Massenez in his paper read before the institute.
The metal, after three minutes' after-blow, gave : phosphorus ·13 ;
and a further 25 seconds gave phosphorus ·10 ; carbon, a trace ;
manganese, ·17 ; sulphur, ·12. At this stage of the operation a
large quantity of slag was poured out of the converter, and then
the spiegel was added. The steel contained : carbon, ·19 ;
manganese, ·57 ; sulphur, ·10 ; phosphorus, ·10. The steel
worked well under the steam-hammer. The slag was of the
following composition :—iron, 10·20 ; lime, 46·94 ; silica, 9·67 ;
phosphoric acid, 9·70. On Thursday, the 26th August, I visited
the Rhenish Steel Works with several members of the Iron and
Steel Institute, and the samples were analysed by Mr. Cook, who
shows the re-melted metal to contain : combined carbon, 2·90 ;
manganese, 1·10 ; silicon, ·46 ; sulphur, ·16 ; phosphorus, 2·03.

The after-blow was very long, being nearly $4\frac{1}{4}$ minutes before the first sample was taken, and a further $\frac{3}{4}$ minute before the second sample was taken, in all five minutes. The carbon lines appeared on the spectroscope in a few seconds after the converter was turned up. The steel contained: carbon, ·28 ; manganese, ·56 ; sulphur, ·08 ; phosphorus, ·08. The metal before the addition of the spiegel had phosphorus, ·07. The slag here is not poured off before the spiegel is added. The sample of slag analysed by Mr. Cook is almost identically the same as that given above from the Hôerde Works. Another cast made when about 150 members of the institute were present contained, I am informed, phosphorus ·13. It was most difficult to get near the workmen who were testing the samples, so great was the crush and the desire to obtain a piece of the metal ; and the wonder was that the metal was so well blown and so low in phosphorus, considering the circumstances under which the operation was performed.

" Messrs. Bolckow, Vaughan, and Co. resolved to erect some large converters at the Cleveland Steel Works, of a size and form which they expected would enable them to overcome some of the difficulties which they had experienced when working with the old converters on the basic system. This new form of converter is concentric, whilst the old converters are eccentric. During the operation of blowing, the lime and metal are lighted by the force of the blast, and when that force is somewhat expended the materials fall again on to the bottom in the new form, whilst in the old form some portions would cling to the nose. The concentric form has also another advantage ; it gives a much larger area of floor to work in, by enabling the metal to be poured into the converter when turned on its side with its nose pointing away from the converter-ladle crane, just the contrary of the present practice. On the 18th October, 1880, this converter was set to work on the basic system, and was quite successful, answering the purpose well, and showing no more symptoms of gathering at the outlet than when making ordinary steel. Our plan of operations is exceedingly simple. The converter, as is usual, is first heated up with coke, so as to prevent the chilling of the metal; then a measured quantity of well-burnt lime, about 16 per cent. of the weight of molten metal, mixed with a small quantity of coal and coke, is charged into the converter, and blown till the lime is well heated. The molten metal is then poured on the lime additions, the blast, 25 lb. pressure, is turned on, and the carbon lines disappear in about ten minutes; then after about two and a half minutes' over-blow, the converter is turned down, and a small sample ingot made, which is quickly beaten into a thin sheet under a small steam-hammer, cooled in water, broken in two pieces, and the fracture shows to the experienced eye whether the metal is sufficiently ductile. If it is not so, then the blowing is prolonged, after which the spiegel is added, and is now being poured into the ladle, not into the converter. For the basic process the metal bath should be low in silicon, because silicon fluxes and destroys the lining, and causes waste of metal; it should be low in sulphur so that the metal may not be red-short. Nearly one-half the sulphur is eliminated by the basic process. In order to work economically, the metal should be taken direct from the blast furnace, so as to avoid—(1) the cost of re-melting in a cupola, and (2) to avoid further contact of the metal with the sulphur and the impurities of the coke. It is not an easy matter to accomplish in a blast furnace the manufacture of a metal low in silicon and at the same time low in sulphur. It would, no doubt, very much help to keep sulphur low if manganese were used, but manganese is a costly metal. At present we have succeeded in making a mottled Cleveland iron with 1 per cent. of silicon and ·16 per cent. sulphur, and white iron with ·5 per cent. silicon and ·25 per cent. sulphur, which, taken direct from the blast furnace, have both made excellent steel; but we have another method of operating, which relieves us from the necessity of making a particular quality of Cleveland pig iron. We call this second mode of working the "transfer" system, because we transfer the metal from the acid to the basic converter. The "transfer" system enables us to take any gray iron direct from the blast furnace to the converter, without any consideration as to the percentage of sulphur, which is always low in gray iron. This gray metal is poured into a converter with a silicious lining, and desiliconised, when, after, say, twelve or fifteen minutes' blowing in the ordinary manner, it is poured out of the converter into the ladle, and poured again from that ladle into a converter lined with dolomite, taking care that the highly silicious slag is prevented from entering the basic-lined converter. Then in the second converter it is only necessary to add sufficient lime for the absorption of the phosphorus of the metal, and the blowing then used need not occupy more time than is necessary for the elimination of the phosphorus—say, about three minutes. This mode of operation will, no doubt, give the basic lining and bottom a much longer life; but both systems are good and effective, and have given excellent results. I have thus summed up in ten minutes what has taken about two years of constant work and the expenditure of large sums of money to accomplish. I am now able to say that the basic process has been brought to a technical and commercial success at the Cleveland Steel Works of Messrs. Bolckow, Vaughan, and Co.

" One feature in this new process seems to have been lost sight of by those who have written on the subject—namely, the possibility or otherwise of being able to eliminate phosphorus before the carbon flame drops, so as to avoid the after-blow. Few give any hope of this being accomplished; but when we remember that few gave any hope of the basic process, or of any other process being successful in eliminating phosphorus at the high temperature of the Bessemer converter, we should not abandon research or relax efforts. It has been said over and over again that the basic process was a failure, and would never succeed. It is a grand trait in the character of Englishmen—that of not knowing when they are beaten. If the after-blow could be avoided, the wear of the lining and bottoms would be very much reduced. We know already that the basic lining will not be anything like so enduring as the acid lining, so special means have been adopted to quickly change a converter. An overhead steam travelling crane capable of lifting 60 tons is being erected, so that, directly a converter lining has worn out, the crane will remove the worn converter out of the way, and bring in a re-lined one dried and ready for working. A very ingenious plan for quickly changing the converter without removing the trunnion was patented by Mr. Holley, the well-known American engineer and metallurgist."

Such was the rapidity with which the mechanical difficulties of the process were overcome that Mr. Holley stated, in 1881, that the manufacture of steel in basic-lined converters was as far advanced as the older alternative process was five years previously. The progress of the process has been more rapid than that of any other great invention in the trade; and next to the invention of the Bessemer converter, it is likely to cause the greatest revolution in the means available for the conversion of iron into steel. It was soon adopted in Austria, Belgium, France, Germany, England, and the United States. Even the Bessemer process did not make anything like such rapid progress in the first years of its existence, probably owing to its greater novelty.

As an illustration of the eagerness of Continental ironmasters to avail themselves of the new process, we may mention the following amusing circumstance :—A Continental ironmaster called on the patentee one April morning at the unusual hour of half-past seven, and at once proceeded to negotiate terms for the use of the process. They were engaged together in conversation for three hours arranging matters of detail, and just as they had concluded, a telegram arrived announcing to the patentee that another Continental ironmaster from the same district was coming to arrange terms for acquiring the same rights in order to work the process. The first visitor, however, had secured a monopoly of the patent rights for that district ; and the second, on his arrival at noon, found that he was too late. It afterwards transpired that both these ironmasters had come over in the same boat to London, and that the one on landing drove direct to the house of the patentee, while the other went to a hotel to rest himself and get breakfast before entering upon the business of his mission.

Other German manufacturers tried to work the process regardless of patent rights. The validity of the patents was attacked by a powerful combination of the North German steel manufacturers, who, by taking advantage of some irregularity and delay in the proceedings connected with the procuring of the patents, and by simultaneously challenging them on various other grounds, endeavoured to free themselves from the obligation of paying royalties for an invention which was acknowledged by them to be of the greatest importance. In 1880 two cases were tried at Berlin, and the judges included the head of the Imperial Patent Office and other eminent jurists, as well as Professor Wedding, Dr. Bruno Kerl, Dr. Hoffman, Dr. Siemens, and other technical authorities. In both cases the courts held the validity of the patents to be thoroughly established, and considered the substantial novelty and great value of the invention to be proved, and to be such as to amply cover any minor technical defects. This decision was generally welcomed as showing that the German Patent Court was determined to administer the new law on just and equitable principles, and not on the narrow basis of the old law, which refused protection to the inventions of Bessemer and Siemens.

As further evidence of its appreciation on the Continent, it may be mentioned that early in 1880 the Westphalian ironmasters offered 150,000*l.* for the acquisition of the patent rights in that district ; but the offer was declined. In 1881, when Mr. Thomas was travelling in the United States, where he received a cordial welcome, the American steel manufacturers offered 55,000*l.* for the use of the patent in that country, which needs it less perhaps than any other country. The offer was accepted.

In one respect its history differs from that of the Bessemer converter. Sir Henry Bessemer's claims to the honour of being the first inventor of the converter were never seriously disputed ; but the young inventors of the basic process, like many great inventors before them, have the satisfaction of knowing that some of the greatest metallurgists in England and on the Continent now claim to have previously discovered the principle of their process. To them, however, belongs the honour of being the first to establish a practical and cheap process of dephosphori-sation ; and that, too, at a time when the iron trade was despairing of such a consummation, which it had vainly hoped for so long. Its ultimate effect need not be anticipated. Its present effect has been to create almost a new industry in some of the greatest industrial centres of the Continent; and in England its general adoption would enable her to multiply her means of production fourfold. The greatest metallurgists both in England and on the Continent have pronounced the steel produced by it to be of high quality, in no respect different from hematite steel, and especially adapted for rails, plates, and other industrial purposes ; indeed, it is held to be superior to hematite steel for certain purposes, such as wire, boiler and other plates. So great was the impression that it made on the Continent that new works were quickly built expressly for the conversion of the poorest ores into steel. Before the process had been three years in operation it was the means of producing nearly half a million tons of steel per annum.

In 1883 the Iron and Steel Institute resolved to award the Bessemer gold medal to Mr. Thomas in recognition of the value of his invention to the steel trade of the world ; but as he was then in Australia, which he visited for the benefit of his health, the presentation was deferred at the request of his mother till his return home. Mr. Thomas returned in the summer; but in the autumn again left England for Algeria. In a letter to the President of the Institute (Mr. B. Samuelson, M.P.), acknowledging his appreciation of the honour conferred on him, he said : " It would be difficult for me to insist too strongly on how greatly we are indebted for the success the basic process has now attained to the unwearied exertions, the conspicuous energy and ability of my colleague, Mr. Gilchrist, whom I regard as no less my associate in the acceptance of this medal than he was in the sometimes anxious days of which this is the outcome. I am sure, too, that he and I are agreed in saying that the present position of dephosphorisation has been only rendered possible by the frank, generous, and unreserved co-operation of Mr. Richards. As an instance of the effect of free discussion of metallurgical theories and experience, which this institute especially promotes, it may be interesting to note that, while in the autumn of 1877 there was, so far as I know, no public record of even any successful experiment tending to show that phosphorus could be removed in the Bessemer or Siemens process, for the present month of September (1883) the make of dephosphorised Bessemer and Siemens steel is between sixty and seventy thousand tons."

MR. GEORGE JAMES SNELUS.

CHAPTER XI.

" Every positive determination in science is susceptible of extension and of useful application, though the period may be distant. A microscopical observation or an optical property, which at first sight is only curious and abstract, may in time become important to agriculturists and manufacturers, —BIOT.

THERE are four great inventions or discoveries that have given a great impetus to the manufacture of iron and steel during the present century, namely, the introduction of hot blast into the blast furnace for the production of crude iron ; the application of the cold blast in the Bessemer converter for the conversion of liquid iron into steel ; the production of steel direct from the ore on the open hearth ; and the discovery of a basic lining by which phosphorus is eliminated and all qualities of iron converted into steel. It is remarkable that only one of the inventors of these im provements was directly or professionally connected with the iron trade. The discovery, says Dr. Ure, of the superior power of a hot over a cold blast in fusing refractory lumps of cast-iron was accidentally made by my pupil, Mr. James Beaumont Neilson, engineer to the Glasgow Gas Works, about the year 1827, at a smith's forge in that city, and it was made the subject of a patent in the month of September of the following year. Sir Henry Bessemer, who thirty years afterwards invented his converter for the production of steel by the application of cold blast, was also an engineer, who three years previously had no knowledge of the iron trade. Sir William Siemens, who both designed and put in operation the direct process of producing steel from raw ores, was an engineer and electrician likewise unconnected with the iron trade; and Mr. Thomas, whose name has been most prominently associated with the basic process of dephosphorisation, was a member of the Civil Service previous to his labours in perfecting that process. To Mr. Snelus belongs the honour of having been the first to discover the principle of the basic process, and he is the only man connected with the trade whose investigations have been attended with such pregnant results as entitle him to a place among its successful inventors. No one in the steel trade has had a more distinguished career as a metallurgist, and perhaps no one has been more scientific in his data or more accurate in his conclusions. He is a recognised authority on both practical and scientific questions in connection with the trade.

George James Snelus was born at Camden Town, London, on June 24, 1837. His father, James Snelus, a builder, died before young George had reached his twelfth year. The son was nevertheless well educated, and trained as a teacher; but he had a preference for applied science, which an accidental circumstance turned into the study of chemistry and metallurgy. Shortly after the invention of the Bessemer process, he happened to attend a lecture on it at the Polytechnic by Professor Pepper, and was so fascinated with the subject that he determined to devote himself to applied science. After some preliminary instruction, he became a student under Professor Roscoe at Owens College in Manchester. He thus qualified himself as a teacher of science under the Department of Science and Art; and in order to complete his scientific training, he next became a student at the Royal School of Mines. Accordingly he studied there from 1864 to 1867; and his career was one of pre-eminent distinction. In the science examination of May, 1864, he gained the Royal Albert scholarship; he also took the first place and the gold medal for physical geography ; the first place and the silver medal for applied mechanics; the second place and the bronze medal for inorganic chemistry; the third place and the bronze medal for magnetism and electricity; and lastly, the De la Beche medal for mining. During the second year of his studies at the School of Mines he became the assistant of Professor Frankland at the Royal College of Chemistry, and successfully conducted science classes at the Royal Polytechnic and other institutions. At the end of his curriculum at the School of Mines he became an Associate in Mining and Metallurgy. He was then (1867) appointed chemist at the Dowlais Iron and Steel Works, which were at that time the largest in the United Kingdom.

In these busy years his name was frequently conspicuous in connection with the rifle volunteer movement. He became a volunteer in 1860, and besides being for many years a commissioned officer, he occupied a foremost place as a successful "shot." In 1864 he won a place in the first 60 for the Queen's Prize at Wimbledon; in 1868 he won the first prize for small-bore rifles in the Albert competition; and he frequently won other prizes. For twelve years—1866-77—he was one of the competitors in the volunteer match between England, Ireland, and Scotland, and his name often stood among the highest scores.

Mr. Snelus was one of the first members of the Iron and Steel Institute, and from its formation has been a frequent contributor and speaker at its meetings. His first paper was read at the Merthyr Tydvil meeting in 1870, " On the condition of carbon and silicon in iron and steel." Its fresh information and elegant composition at once attracted attention ; it combated some of the views on the subject propounded by Dr. Percy and other metallurgists; and its conclusions were adopted by Mr. I. L. Bell. Next year he investigated the composition of the gases evolved from the Bessemer converter during the blow. At the annual meeting of the Iron and Steel Institute held in London in March, 1871, Professor Roscoe delivered a lecture on the spectrum analysis of the flame issuing from the Bessemer converter, and in it he alluded to the difficulty of determining the cause of the greater part of the lines in the Bessemer spectrum, pointing out that while most observers referred them to carbon in some form, others believed them to be mainly due to manganese. Mr. Snelus thought an analysis of the gas producing the lines would be a step towards solving the difficulty. Although it was generally assumed that during the process of conversion the carbon in the iron was burnt to carbonic oxide, that theory had never been proved; and Mr. Snelus perceived that the composition of the gases evolved from the converter would afford an insight into the nature of the process going on inside. Accordingly he collected the gas for analysis by means of a long iron gas-pipe, one end of which was inserted in the neck of the converter, while at the other end glass tubes were attached at particular periods of the blow when the gas was to be analysed, and these were sealed up with the blow-pipe before being removed. As a constant stream of gas was allowed to rush through the pipe, Mr. Snelus felt sure that the gas thus collected was a fair sample of that produced at the time in the converter. It was observable, he says, that during the first part of the blow the gas would not light at the end of the iron tube, but from about the commencement of the " boil " to the end of the blow it burned with the pale blue flame characteristic of carbonic oxide. An analysis of the samples of gas, which were taken every two minutes, showed, broadly speaking, that the carbon in the converter took up twice as much oxygen at one time as at another; and he inferred that this fact and the difference of spectra were due to temperature. It was certain, he said, that at the commencement of the blow the temperature could not be much above a yellow heat, while at the end of the blow it was undoubtedly a good white heat. His analyses also showed that the gas from the Bessemer converter during the last half of the blow is really of as much value as any gas made purposely or incidentally at an iron works. "This," he remarked, "is an important fact and if we consider the amount of fuel thus going to waste, the question naturally suggests itself whether it cannot be made available. If we assume that, during the latter half of the blow, two thirds of the total carbon in the pig is burnt, and take the melted iron to contain three and a half per cent. of carbon, we find that a Bessemer works using only 1,000 tons of such pig per week is wasting a quantity of fuel equal to 23⅓ tons of pure carbon, or, say, 25 tons of coke (40 tons of coal). Now, I would suggest that it is possible by simple mechanism to collect this gas and pass it under boilers, where it would save its equivalent of coal. The large body of flame is not wanted for any purpose. True, the workman now depends upon its indications to afford him the means of judging when the blow is completed, but the spectroscope would do this with greater accuracy with a fraction of the gas which now roars from the converter."

About the time he was pursuing these investigations the question that was exercising the iron trade was whether iron could be puddled or refined in a machine so as to supersede hand labour ; Mr. Danks, an American, had invented a machine for that purpose, and the members of the Iron and Steel Institute were so much interested in the question of its success, that in 1871 they resolved to send a deputation to America to investigate the subject. Mr. Snelus was selected for that purpose as the best authority for solving the scientific questions involved in the process. His collaborateurs were Messrs. Jones and Lester. The Commission sailed from Liverpool early in October, and visited the Cincinnati Iron Works, where the Danks puddling machine was at work, and where they conducted an elaborate series of experiments for the purpose of testing its capabilities. Along with the Commission were sent forty tons of pig-iron selected from Dowlais, Coneygree, Butterley, and Cleveland, together with a variety of fluxing materials, all of which were used in the Danks rotatory furnace. The general opinion in America at that time was that the machine was a success; and high hopes were entertained that at last a mechanical means of refining pig-iron would supersede manual labour. The Commission drew up their report at Washington on December 12, and submitted it for the consideration of the committee of the Iron and Steel Institute on January 12. In addition to his labours in the production of this joint report, Mr. Snelus drew up a special one on the scientific features of the process. The analyses con- tained in that report were the most elaborate that had been made of the puddling process up to that time; and they were of permanent value as revealing the important fact, hitherto unsus- pected, that phosphorus was eliminated very early in the puddling process while the metal was in a fluid condition. Till then Dr. Percy had taught that phosphorus was removed by a process of liquation from the puddled ball; but Mr. Snelus now exploded this theory; and this discovery was the germ of the principle that has since borne such fruitful results in the steel trade.

The Danks puddling process was not the only one he then investigated. At that time there was a kind of mania for puddling processes, and one after another came up for examination. Mr. Snelus did valuable service to the iron trade by the thoroughness and impartiality with which he investigated their scientific features. In this way he examined the Sherman and Heaton processes, both of which attracted attention in their day, and presented the results in reports to the Iron and Steel Institute. In addition to his own reports on these processes, he took a leading part in several of the discussions on other inventions which had the same object in view, but all of which have since been superseded. He also contributed papers on the "Manufacture and use of spiegeleisen;" and on "Fireclays and other refractory materials." The subject matter alone of these papers indicates the range of Mr. Snelus' researches, and all of them were, moreover, distinguished by their original information and lucid arrangement

The first announcements of the principle of the basic process of dephosphorisation form another example of the way in which a valuable discovery is ofttimes allowed to lie unused by its discoverer, till somebody else lays claim to it, and turns it to good account. At the annual meeting of the Iron and Steel Institute held in London in May, 1872, under the presidency of Sir Henry Bessemer, a lengthy discussion took place on the reports presented on the Danks and other puddling processes. In the course of it Mr. Snelus said that no one could be more satisfied than himself of the injurious effects of phosphorus upon steel, and of the fact that it did not go out in the Bessemer process. He had his own opinion about the reasons why it did not go out in the Bessemer process, which, however, he was not then at liberty to make known, because he was working in that direction, and had some hopes of surmounting the difficulty, though he had not then gone far enough with his experiments. Next year these ambiguous allusions were made more explicit.

When Sir William Siemens explained to the Iron and Steel Institute in 1873 the difficulties he had experienced in finding a lining for his steel furnace, that was capable of resisting the excessive heat necessary for carrying out his process, Mr. Snelus stated, in the discussion which followed, that in the previous year he had taken out a patent for using limestone in places where Sir William Siemens was using bauxite, and that in experimenting with it he had used it for the lining of a small Bessemer vessel, in which it stood admirably. The way in which he had used it was by grinding the limestone to a powder, not using raw lime, but taking limestone and grinding it to the plastic condition of ordinary clay. He found it was about as plastic as ordinary clay, and that it could be moulded into any shape. When rammed up round a core, it formed a permanent lining as long as it was in use. Of course, he added, if it had to stand any length of time the lime produced would slake away ; but as long as it was in use—and he saw no reason why they could not keep a furnace of that description constantly in use—the temperature had no action upon it, and it formed a hard, compact, and infusible lining. He thought Sir William Siemens would find it more infusible than bauxite, as well as cheaper and perhaps more practical to use ; but he would leave that for Sir William to experiment upon, and would be very glad to hear the results if that gentleman would give it a trial.

The experiments which he thus clearly indicated were not made known publicly till seven years afterwards, and then they created quite a sensation, for it was made perfectly evident that all that time Mr. Snelus was acquainted with the principle and properties of the basic lining as a means of dephosphorising iron in the Bessemer converter. There are many well attested instances on record in which different individuals have made the same discoveries simultaneously in different localities ; and the present age adds many such instances to the records of the past. When the history of the inventions of the nineteenth century is properly written, it will be found that the conception of some of the most original and useful of them was allowed to lie dormant till the course of events forced their practical application, or till the announcement of some competitor in the same field of inquiry brought forward the original inventor to vindicate his claims. The basic process of dephosphorisation belongs to this category ; but a not less notable feature of its history is the fact that its principle and material were publicly, though incompletely, pointed out to Sir Henry Bessemer and Sir William Siemens by two men of scientific attainments ; and neither of these great inventors utilised the valuable suggestions gratuitously offered to them.

Mr. Snelus was first led to doubt the correctness of Dr. Percy's hypothesis, already referred to, that "phosphorus is eliminated by a process of liquation of fluid phosphide of iron from the pasty puddled ball," during his studies of the puddling and refining processes at Dowlais; and in examining the reactions that occured in the Danks process he was struck by the fact that a good deal of the phosphorus was eliminated while the iron was in a fluid state. Impressed with the importance of this discovery, for such it was, he determined to investigate the matter still further. A comparison of the slags in the process of Welsh refining, puddling, and Bessemer converting showed him that the slag from the puddling process, in which the phosphorus was comparatively easily removed, was highly basic; while that from the refinery, in which only small amounts of phosphorus were removed, was less basic; and that from the converter, in which none at all was removed, was highly silicious. Other observations pointed to the same results, and led Mr. Snelus to the conclusion that phosphorus was removed in all processes just in proportion to the basic nature of the slag. With these data before him, he was meditating, in the beginning of 1872, how he could systematically obtain a highly-basic slag in the Bessemer process; and he tried two or three materials of this nature, but found that they were not sufficiently refractory to withstand the intense heat of the converter. While his experiments were in this tentative state, he happened to make the fortunate observation that lime, when subjected to an excessively high temperature—in the Siemens steel melting furnace —became indurated and insensible to moisture. This fact, accidentally observed while performing an experiment for another purpose, suggested to him that lime in some form or other was the basic material he required, for lining the converter; and being well aware that it would not do to burn lime at too high a temperature if it were to be caustic, he selected and particularly specified magnesian limestone. Shortly afterwards he found that bricks could be made of lime or limestone, if the lime, when used for that purpose, were crushed quickly, compressed, and fired before it had time to absorb moisture. To consolidate these bricks, however, required an intense temperature, the more so when the lime was pure ; but he found that it was facilitated by the application of a small quantity of oxide of iron or other fusible base. In order to put these conclusions to a practical test, he lined a small converter with crushed limestone, and fired it at a very high temperature.

His first experiment was made with one or two cwt. of Cleveland pig, which was melted in a cupola, then poured into this converter, and blown in the usual way till the carbon lines disappeared. The first piece of steel thus produced from phosphoric iron was carefully preserved by Mr. Snelus. The metal was found to be free from phosphorus, which, however, was traced into the slag. Several other blows were made, and samples of metal and slag taken at intervals during the blow. On analyses being made, the results showed clearly that so long as the slag was kept basic there was no difficulty in removing the phosphorus. To make sure that this was the secret of success, he had the same small converter lined with ganister, and then he found that the phosphorus, as hitherto, remained in the metal. Next he tried to line a seven ton converter with the same material that was used in the small one ; but he found considerable difficulty in preparing a lining sufficiently strong of that size. This led him to defer further experiments till a more convenient season ; but in the meantime he secured his discovery by patent. In his patent he specified the use of lime and limestone, magnesian or otherwise, in all the forms he thought it possible to use it, for " the lining of all furnaces in which metals or oxides are melted or operated upon while fluid." " I felt," he says, " that, although I had found the solution of the problem so long and anxiously looked for, my plans required to be tried on a larger scale before they were fully laid before the public, not because there could be any question that the same chemical action that succeeded with one or two cwt. of metal was likely to be reversed when operating upon tons, but because I felt that there were many points of detail that would require to be worked out. These details have been filled up by subsequent workers, but it is gratifying to me that the complete success now attained follows closely some of the lines then laid down. To the question why I did not follow up my discovery and put my plans into practice, my answer is that I had just taken the management of a concern the interests of which were opposed to the solution of this problem, and therefore having secured the ground by patent, I was compelled to wait a more favourable opportunity for putting my plans into practice. There was also a good deal of scepticism among those interested as to the possibility of solving the problem, and none seemed to care to take it up." He tried to get some of his friends to work his process, but in this respect he was unsuccessful.

The correctness of this account is verified by some remarkable facts. Mr. Snelus had confidentially told some of his friends of the experimental success of his invention in the long interval between its conception and perfection. Among them was an eminent engineer, who, on hearing that a patent had been taken out for this process, strongly advised Mr. Snelus not to dabble in patents, as they did not pay; but to show that his mind was open to conviction, this friend added : " when you can show 1,000 tons of rails made by the process, then I will believe in it." In May, 1880, this same friend stated that he had recently had occasion to test 1,000 tons of rails made by the basic process as compared with 1,000 tons made by the old process, and that he found no difference whatever. Again, a few minutes before Mr. Thomas vaguely announced his discovery of the same process and his more perfect application of it, Mr. Snelus gave a short account of its leading features to the Iron and Steel Institute. Before any one else mentioned the elimination of phosphorus by the use of magnesian limestone, Mr. Snelus stated that he had succeeded in that way ; and that, though there was a difficulty in constructing a practical apparatus in which the limestone could be used, he had such confidence in the ultimate success of that material for this purpose, that he still intended to maintain his patent, although it was six years old, and the seven years' fee would soon have to be paid. The story of the eventual application and success of the process has been told in the previous chapter.

The rival inventors and patentees had the good sense to avoid litigation as a means of settling their claims. Messrs Thomas and Gilchrist, who made the process a practical success, agreed with Mr. Snelus to submit their claims as to the way in which the profits should be divided to the decision of Sir William Thomson ; and by this amicable arrangement no legal difficulty was allowed to retard the application of the process. The patent rights of Mr. Snelus extend to America and the United Kingdom ; but not to the Continent. The history of his American patent is another evidence of his priority of invention as well as his dilatoriness in availing himself of it. When he discovered in 1872 that in principle he had solved the problem, he shortly afterwards fully described it to his friend Mr. E. Cooper, of Cooper & Hewitt of New York, and that gentleman urged him at once to take out an American patent; but here again Mr. Snelus' habitual desire to thoroughly perfect all details nearly lost him his rights. In the hope that he would soon find an opportunity of perfecting the working of the process, he disregarded his friend's advice until Mr. Thomas announced his discovery of it in 1878. Mr. Snelus then joined Messrs. Cooper & Hewitt in taking out an American patent, which secured the original inventor a share in the royalties from America.

After the process became a practical success, he applied himself to its elucidation. As the result of careful observations he concluded that the elements found in combination with iron were eliminated in the following order: in a basic lined converter, silicon went first, carbon second, phosphorus next, then manganese, and lastly sulphur. In a converter with the old lining, called, for the sake of distinction, an acid lining, generally silicon went first, then carbon, and lastly manganese, while phosphorus and sulphur were not removed in the slightest degree. No doubt temperature would be found to modify the results in a basic lined converter; and although he believed the combustion of the elements took place in this relative order, yet one reaction to some extent overlapped the other. Thus, in an acid lined converter carbon was gradually removed from the commencement; but while the silicon disappeared rapidly in the early part of the blow, the carbon only went slowly.

As to the ultimate effect of the process, Mr. Snelus believed from the first that it was not only capable of producing from the poorest ores a metal having all the properties of Bessemer steel, but also of making the finest classes of steel for cutlery purposes. After careful investigation he has come to the conclusion that the essential for a steel to carry a keen cutting edge, without being brittle, is that it shall not only be free from silicon and have the right proportion of carbon, but that it shall be absolutely free from phosphorus. He believes that the basic process is capable of producing metal, even from inferior pig, eminently suitable for the manufacture of cutlery and other goods, at one fourth the price of the high-class steel hitherto used for that purpose.

The work which so absorbed Mr. Snelus' attention as to put the basic process in abeyance was the remodelling of the West Cumberland Steel Works. His appointment as manager of the Bessemer department of these works occurred shortly after he had taken out his patent in England; and a few months afterwards he was promoted to the position of general manager of the entire works. At the time this promotion took place he was at a loss to know which would be the best way to perfect the basic process before making it thoroughly public, and more urgent affairs soon absorbed his powers. "My duties at West Cumberland," he says, "demanded all my time and attention, and I had to postpone further trials till a more convenient occasion. Several years of anxious and hard work then intervened, and being determined to make a success of the West Cumberland Works, I had to devote all my time and attention to them." Under his management these works did become a success. His scientific and practical knowledge of the Bessemer process enabled him to introduce some valuable improvements. Of his scientific attainments we have already given some account; his practical knowledge was also extensive. During his visit to America he examined most of the large works there, and took notes of everything he saw in connection with the manufacture of iron and steel. He made careful observations of the improvements made there in the mechanical details of the Bessemer process; and was so impressed with the advantages of the Holley patent for Bessemer vessel bottoms, that he made a special journey of 1,000 miles, from New York to Harrisburg, in order to satisfy himself at the latter place that he perfectly understood the details of this invention. He afterwards introduced the Holley patent at the West Cumberland Works, where it continued in use. The year after his return from America he visited Germany, and he made a sojourn at Hôerde purposely to see the process of ascensional casting under Mr. Pink. He was so convinced of its utility that he wished to introduce this improvement also at West Cumberland, but found that the size and construction of the Bessemer pits there were not suitable for it.

His next improvement at West Cumberland was of a more interesting character. At that time the introduction of the fluid metal direct from the blast furnace to the Bessemer converter was engaging the attention of steel makers. When Sir Henry Bessemer first publicly described the details of his process, he

stated that the fluid crude iron was to be run into the converter
from the blast furnace ; and that, after " the blow," the ingots
were to be put on a truck and run along a line of rails to the
rolls, the whole of the latter operation lasting a minute or two,
" in which time," he said, " the ingots will not have cooled down
sufficiently in the centre for rolling, so that the first bars will be
produced and finished off fit for sale wholly without the use
of fuel, and within a period of forty minutes from the time of
tapping the crude iron from the blast furnace. If the iron is
made in very large quantities, the heat of some of the ingots
will not be retained until they can be rolled into bars ; a small
oven, capable of retaining the heat, must, in that case, be erected
near the rolls, in which the ingots may remain till the rolls are
at liberty." Nearly a quarter of a century elapsed before these
directions were carried out ; and when carried out, the one was
considered a new discovery, and the other was heralded as a great
invention. When the Bessemer process was first put in operation
in Sheffield, the iron used in it was manufactured either in
Sweden or West Cumberland, and consequently it had to be
remelted, after transit, in order to be " converted." In course
of time this mere accident of circumstances came to be regarded
as an essential principle in English metallurgy ; and when in
1874-5 it was reported in England that at some foreign works the
molten iron was being run direct from the blast furnace into the
converter, it was contended by many that to convert the fluid
metal from the blast furnace into steel, without first cooling
and then remelting it, would be impracticable in working and
uncertain in results.

Mr. Snelus was among the first in England to advocate the
conversion of the fluid iron direct from the blast furnace into
steel. For carrying out this improvement he designed appliances
of a model kind. At the West Cumberland Iron and Steel
Works the distance between the blast furnaces and the converters

was 350 yards. To carry the molten metal this distance he designed a large ladle capable of holding 8 tons. It was so placed in the centre of a carriage that the greatest stability was obtained without the need of any balance weight, and it was turned over by a simple cast steel worm and wheel. The weight of the whole apparatus was under 10 tons. In practice about 3 tons 10 cwt. of molten iron was tapped from each of two furnaces into the ladle in order to ensure a uniform charge. The tapping of both furnaces was often done in about five minutes, and in less than that time the charge was weighed, taken to the converter, and poured into it. One ladle lasted from 100 to 200 casts before the "skull" needed to be taken out. By this method considerable benefit arose from the increased yield of metal; and Mr. Snelus stated in 1876 that he estimated the saving in labour and fuel at 4*s.* or 5*s.* per ton. He also found that the steel made in this way was better in quality.

The advent of another improvement forms an episode in the history of the Bessemer process. In the autumn of 1882 Mr. Gjers of Middlesbrough, in introducing an invention of his to the notice of the iron and steel trade, stated that "when Sir Henry Bessemer in 1856 made public his great invention, and announced to the world that he was able to produce malleable steel from cast iron without the expenditure of any fuel, except that which already existed in the fluid metal, imparted to it in the blast furnace, his statement was received with doubt and surprise. If at that time he had been able to add that it was also possible to roll such steel into a finished bar with no further expenditure of fuel, then, undoubtedly, the surprise would have been still greater. Even this, however, has come to pass, and it is now easy and practicable to roll a bloom into a rail or other finished article with its own initial heat." Perhaps the most surprising thing in connection with this idea is the fact that Sir Henry Bessemer was the first to suggest and announce it, as appears from the extract just given from the paper he read at Cheltenham in 1856. In the intervening quarter of a century, the universal practice was to put the ingots into a heating furnace in order to impart to them the high uniform temperature at which they could be rolled; so that when, in 1882, Mr. Gjers published his design of "an oven" composed of cells of brickwork, called by him "soaking pits," because the ingots in them were soaked in their own heat till they were fit to be rolled, he was hailed as a public benefactor, and his invention was described as being both economically advantageous and as simplifying the operation of steel making on a large scale. It was said to dispense with a large number of men—some of them highly paid—in connection with the heating department; to do away with costly heating furnaces and gas generators; to save all the coal used in heating; and to save the loss in yield of steel, as there would be no more steel spoiled by overheating in the furnaces. Mr. Snelus was one of the first steel makers in England to adopt this improvement at the West Cumberland Works; and when, for the first time, the subject was brought before the Iron and Steel Institute at Vienna in September, 1882, he spoke strongly in favour of it, and gave some practical advice as to the construction of "soaking pits."

At the Paris Exhibition of 1878 he was an exhibitor on his own account of a very elaborate collection of analysed samples illustrating the manufacture of iron and steel in England. For this the jury awarded him a gold medal as a "collaborateur." The collection was purchased by Professor Deure for the Polytechnic School at Aix-la-Chapelle.

In 1883 the President of the Iron and Steel Institute (Mr. B. Samuelson, M.P.) presented the Bessemer gold medal to Mr. Snelus "as the first man who made pure steel from impure iron in a Bessemer converter lined with basic materials." Mr. Snelus' reply was interesting and graceful. He said: "It is not a little singular that my first lesson in metallurgy and science was obtained at a lecture by Professor Pepper on the Bessemer process. The clear and lucid way in which he presented the great discovery of Sir Henry Bessemer quite enchanted me, and I have never ceased to look upon the Bessemer process as one of the most charming phenomena that any man could witness. This being the case, it will not be considered singular that, after having passed my course of studies with Professor Roscoe at Owens College, and under a professor of the School of Mines, I was selected by Dr. Percy to be the chemist to the great Dowlais Works. I thought I had a great future before me. I entered upon those duties with great pleasure, but I met there with such cordial kindness at the hands of the late lamented Mr. Menelaus, that I cannot refrain from mentioning the great assistance which I received from him. It was to his kindly, genial help that I owed a great deal, and it was at Dowlais that I worked out the theory of the basic process. I had the good fortune to be selected by this Institute to report upon the chemistry of Danks's process, thus enabling me in pursuing my investigations to discover the fact, until then unknown, that phosphorus could be removed from pig iron while the pig iron was in the molten state. A discussion with Mr. Isaac Lowthian Bell, after my return from America, induced me to make an experiment, which showed me the simple fact that lime could be burnt to such a high temperature as to be impervious to water. It at once struck me that I had here the means of lining the Bessemer converter so as to be able to blow the metal in it and yet retain the basic lining. I thought it advisable at this early stage to take a provisional patent of what was then an idea, but shortly afterwards I had the opportunity of proving my theory by a course of very severe and careful inspection. Commencing with $1\frac{1}{2}$ to 2 cwt. of Middlesbrough pig iron, I ended by blowing 3 tons in an ordinary Bessemer converter, and in every case I succeeded in removing the phosphorus from the pig iron, and in recovering it in the slag. I lacked, however, a good binding material for the basic lining, and this has been supplied by my colleague Mr. Riley. Mr. Sidney Thomas shortly afterwards, with very much more energy than I had shown, followed in the same line, and Mr. Gilchrist and he developed the process of making basic bricks on a large scale. After this he demonstrated much more publicly than I had done the theory of the basic process, and he induced Mr. Windsor Richards to take it up. It was a piece of very good fortune, I consider, that Mr. Thomas succeeded in enlisting the sympathy of Mr. Richards; this was due to Mr. Thomas's perseverance and his determination to make the process public, and to make it go; but I doubt very much whether Mr. Thomas's perseverance and all our skill would have placed the process in the position in which it now is, if it had not been for the indomitable pluck and energy of Mr. Richards, backed by the great power and interest of a large concern like Messrs. Bolckow, Vaughan, and Co. In thanking you for the very kind way in which you have recognised the portion which I have had in this improvement in the iron and steel trade, I can only say that I believe your action in awarding the Bessemer medal, irrespective of all legal technicalities, will have a considerable effect in encouraging those students of science who are pursuing their investigations after truth, and those engineers who are endeavouring to apply the truths so established for the benefit of their fellow men."

INDEX.

z

LONDON : R. CLAY, SONS, AND TAYLOR, PRINTERS.